Praise for

The Joy of Missing Out

Engaging, thought-provoking and delightfully easy to read, *The Joy of Missing Out* provides practical insights and much-needed hope for an overwhelmed society.

—Dr. Susan Biali, M.D., Psychology Today blogger
and author, *Live a Life You Love*

Christina Crook has crafted a well-researched, utterly readable field guide to finding quiet, and ultimately, ourselves, within today's electronic cacophony.

—Mike Tennant, co-author, *The Age of Persuasion*

Christina Crook writes prophetically about our complicated, often troubled, relationship with technology. But rather than forecasting gloom and doom for the modern age, Crook proposes meaningful questions for cultivating meaningful lives in the present tense. *The Joy of Missing Out* reclaims sacred space for all things blessedly human, and I highly recommend it to readers.

—Jen Pollock Michel, author, *Teach Us to Want*

A skillful meditation on the dynamic of InfoTech and presence. A critique of digital assumptions. A welcome reminder of the real world and its tech shadows.

—Raffi Cavoukian, C.M., singer, author of *Lightweb Darkweb: Three Reasons to Reform Social Media Before It Re-forms Us*

Weaving together research, personal reflections, and philosophy, Crook offers hope that we can create a new relationship with the digital world ... *The Joy of Missing Out* offers practical, tangible advice for ensuring your online habits are aligned with your values.

—Kim Sedgewick, co-founder, Red Tent Sisters

Crook provides both the proof and the inspiration to invest in real-life experiences and establish healthy tech-life balance. ... This book is a life-changer for anyone experiencing the pressure and disconnection of a fast-paced, media-saturated culture.

—Rachel Macy Stafford, New York Times bestselling author,
Hands Free Mama

[Christina] Crook writes a book in search of those human qualities that matter most. In its own way, it is our age's sequel to *The Joy of Cooking* and *The Joy of Sex*. Who knew you could gain so much by selectively giving up so little?

—Dr. Read Mercer Schuchardt, Associate Professor of Communication, Wheaton College

Well-researched and beautifully written, *The Joy of Missing Out* will encourage you to cultivate your attention span, put pen to paper, spend mindful time with your family and rediscover the pleasure of life, unplugged.

—Janine Vangool, publisher and editor, *UPPERCASE* magazine

Crook writes in a way that gives us hope and reminds us of the beauty of "real life", without diminishing the value of technology. Her book is relatable, never preachy; it is finely drawn from experience and research. We can learn from this.

—Dr. Laurie Petrou, Associate Professor, RTA School of Media, Ryerson University

If you've ever found yourself checking your cell phone instead of playing with your children, or surfing the web instead of talking to your best friend, this is the book for you.

—Christopher Meades, author, *The Last Hiccup*

Christina [Crook] calls us to be present with ourselves. Any parent, anyone working in the marketing industry needs to fasten themselves down and read this book (without checking their Facebook newsfeed) and allow Christina to call you to a better place of being. This book is a must-read for anyone who owns a smartphone, has access to the internet, Facebook, Instagram and Twitter.

—Darian Kovacs, Founder and Principal, Jelly Marketing

THE **JOY** OF MISSING OUT

FINDING BALANCE IN A WIRED WORLD

CHRISTINA CROOK

Cover design by Diane McIntosh.
All images © iStock (Meadow: Volokhatiuk; Tablet: studiobra; Wooden Table: pavlen)
Sidebar image © iStock tuna tirkaz

Printed in Canada. First printing February, 2015.

Permissions: A version of the section "Watchful Reverence" (pp 139–140) originally appeared in Issue 10 of *UPPERCASE* magazine, and is used here with kind permission.

A version of "Presentness" (pp 65–8) originally appeared in *The Curator* and is used here with kind permission.

New Society Publishers acknowledges the financial support of the Government of Canada through the Canada Book Fund (CBF) for our publishing activities.

Paperback ISBN: 978-0-86571-767-1 eISBN: 978-1-55092-572-2

Inquiries regarding requests to reprint all or part of The Joy of Missing Out should be addressed to New Society Publishers at the address below.

To order directly from the publishers, please call toll-free (North America) 1-800-567-6772, or order online at www.newsociety.com

Any other inquiries can be directed by mail to:
New Society Publishers
P.O. Box 189, Gabriola Island, BC V0R 1X0, Canada
(250) 247-9737

New Society Publishers' mission is to publish books that contribute in fundamental ways to building an ecologically sustainable and just society, and to do so with the least possible impact on the environment, in a manner that models this vision. We are committed to doing this not just through education, but through action. The interior pages of our bound books are printed on Forest Stewardship Council®-registered acid-free paper that is **100% post-consumer recycled** (100% old growth forest-free), processed chlorine-free, and printed with vegetable-based, low-VOC inks, with covers produced using FSC®-registered stock. New Society also works to reduce its carbon footprint, and purchases carbon offsets based on an annual audit to ensure a carbon neutral footprint. For further information, or to browse our full list of books and purchase securely, visit our website at: www.newsociety.com

LIBRARY AND ARCHIVES CANADA CATALOGUING IN PUBLICATION

Crook, Christina, author
The joy of missing out : finding balance in a wired world / Christina Crook.

Includes bibliographical references.
Issued in print and electronic formats.
ISBN 978-0-86571-767-1 (pbk.).--ISBN 978-1-55092-572-2 (ebook)

1. Conduct of life. 2. Quality of life. 3. Time management. 4. Internet.
5. Technology. I. Title.

BF637.C5C76 2015 158.1 C2014-907560-X
C2014-907561-8

Contents

Introduction

Beware the barrenness of a busy life.

— Socrates

LET ME BEGIN WITH A STORY.

In a darkened room, a small family gathers. The house is silent except for the padding of stocking feet, the murmur of voices. They sit around a small collection of candles while their father strikes a match. The children's eyes dance with wonder at the flame. Normally, this family would flip a switch. In electric-lit halls, they'd walk with intention, but this night they were left without light.

It was their first power outage. No anomaly in this city. *A storm was coming,* the neighbors had said. The parents had laughed it off, standing on the deck hours earlier. They knew the evening's calm. This night would be no different.

They'd just put their youngest child to bed when the house surged into a liquid lull, blackness filling the cavernous living room like a pot of coffee poured out cold. They went in search of the lone flashlight, the one plugged in the wall near the sliding back door. The outward glow of the moon seeped in like a hum and led their hands to the handheld device saved just for such an occasion.

Soon, neighbors began their rounds, knocking on doors, trading candles and beers. Little clusters of candlelight sprung up on front steps. Cheer and laughter spilled out of doors, across darkened pavement, well into the night.

In a town a few miles away, another family is readying for bed. Lantern light fills the rooms as they go about their nighttime routine. Each person steps to the sink, brushes their teeth, scrubs their face. Singing their evening songs, they follow the lantern up stairs to change into their pajamas, though their feet know the steps by rote. It's too dark to read the storybooks now, but the shrieks, laughter and conversations continue as thick blankets are pulled back and little bodies scamper into bed. Parents sing prayers and cuddle up with the youngest as they fall asleep. Day has ended, the fire downstairs is dimming, and soon all will be quiet. Husband and wife gather close in their bed. Tomorrow, at dawn, the day will begin again with the same rhythms.

Back in the city, the family wakes to the groans of a garbage truck lumbering by. Otherwise, it's eerily quiet: no music, no coffee brewing. The old floor boards creak with every step as the first member begins their descent to the living room. Soon everyone is up.

A breakfast of yogurt and cereal is cobbled together in ceramic bowls. Everyone bundles up into snow suits and spends the morning climbing snow banks and careening down on sleds. Afterwards, hands wrap around mugs of hot chocolate (as luck would have it, the gas stove is still working), and dusty board games are tugged down from the shelf. The last phone screen goes black as the remaining power drains out.

The day moves slowly, until it is evening again and everyone is back in their pajamas gathered on the living room floor. The four-year-old daughter pipes up and says what everyone else is thinking:

"Today was a really good day."

A Really Good Day

Kids experience joy all of the time, without knowing it. "Children are often in a state of joy, and it's because they're present, they're living in the moment, they're not focused on their worry about the future or concerns about the past. They're enjoying their moment now," says Dr. Joti Samra, a professor of psychology at Simon Fraser University in Vancouver, BC.

Statistics tell us that most of us are not living with this kind of joy; we are not living our very good days. Instead, we are living in silos. In rooms all over the city, we are gathered side-by-side in cubicles, high above spider-legged byways, spending the currency of our lives in front of MacBook Pros with retina displays, iPads and Androids. Little "poolunk" sounds indicate messages received, interrupting conversations, thoughts and feelings. Neighbors are tucked indoors. Our energies, creativity and time — perhaps the best of us — are being spent committed to screens. Already our gadgets are wearable and, sooner than we think, they'll be under our skin.

Our world is aflutter with a kind of technological mysticism.

> "*New is better.* But these technologies come with an onslaught of unintended consequences.
>
> *Easy is better.* But as machines do more work for us, we do less; we're less capable on our own.
>
> *More is better.* But as machines store and organize more, we get sloppy, forget our friend's phone number, birthday, heartfelt concerns.
>
> *Faster is better.* But as machines enter our way of thinking, we bias speed itself; we lose our capacity for patience. Forget things take time."
>
> — *Geez* magazine, Issue 20, Winter 2010

We weren't born with smartphones in our hands, and we won't be tweeting our own death notices, yet these are the items that dominate our every moment.

Dr. Read Schuchardt, a media ecologist from New York University, explains our digital compulsions this way: "It is very difficult to step out of the immediacy, the 'necessity' of media and say 'maybe I don't need this' because we believe we have control over their effects because we made these technologies. But the truth is, we make our technologies and they remake us in their image and for their purposes."

We've become accustomed to a new way of being *alone together,* says early Internet champion Sherry Turkle, now a growing skeptic. "Technology-enabled, we are able to be with one another, and also elsewhere, connected to wherever we want to be. We want to customize our lives. We want to move in and out of where we are because the thing we value most is control over where we focus our attention."

In my own life, I wanted to untangle the web of my online engagements, so I gave up the Internet for 31 days. I was tired of Facebook mediating my relationships and discontented with my compulsion to constantly check-in online. I knew the Internet was allowing me to emotionally disengage from myself and my loved ones. I was living in a constant state of information overload and a vacuum of joy. I had too much information and not enough wonder. I was seeking beyond what Sherry Turkle calls "our steady state of distracted connectedness."

My "fast" consisted of disabling data on my smartphone and completely turning off my email. I chronicled my experience with a letter a day, complete with news clippings, quotes and thoughts on technology. Each letter was hand- or type-written, mailed, then scanned and posted to a blog by my friend, Marisa Ducklow, creating a conversation between friends and open to the world at large. These letters (some of which are included) set out a narrative that examines the implications, both good and bad, of a technologically focused life. The experience, chronicled as the project, *Letters from a Luddite,* fueled my passion for exploring the intersection of technology, relationships and joy and led to the writing of this book.

The Joy of Missing Out examines the implications of a technologically focused life and the dynamic possibilities for those longing to cultivate a richer on- and off-line existence. Using historical data, type-written letters, "Chapter Challenges," and personal accounts, I make the case for increasing the intentionality in our day-to-day lives, offering solutions for living in a wired world.

This book is divided into three parts. The first section examines the issues, focusing on the rise of passivity, isolation and increased cultural anxiety. It explores the progression from electricity and the telegraph to instantaneous communication — and the massive global shifts in the way we interact with one another.

So, in the first section, we step back to put our lives in context.

The second part focuses on *presentness* and what we can expect to find when we do a fast from the Internet, finding new rhythms in our online pursuits. It explores the writing and research on "getting things done," the impacts of implementing constraints on technology, the outputs of less "connected" individuals, and the impacts of our daily Internet consumption.

The third part of this book considers solutions for living in a wired world. It reveals how key shifts in our thinking can enable us to draw closer to one another, taking up the good burdens of local work and responsibilities. It explores the value of focus, the necessity of viewing the Web as a tool, and the meaning we find in more limited connections. It reconsiders the Western values of power, control and success, revealing how wonder, trust and discipline are central to the experience of being human and the keys to our joy.

By understanding our online habits, we can form new ones — as we seek to be fully human in a smartphone world.

There is life beyond the silo. You can find it. Let this book be your guide.

1

Personhood

The Greatest Tablet in the World

Out of the crooked timber of humanity,
nothing entirely straight can be built.

— Kant

MY EYES ARE BRIGHT WITH READINESS.

I hoist myself upon the metal frame, balancing as I locate the pedals beneath my feet, readying for the open road. I've waited for this ride for days, years. It has long been a dream of mine to pedal a basket-adorned bicycle down a long country road, and today is the culmination of this small, yet urgent dream.

I climb on. Steady myself. Sneakers resting firmly as my hands close in around the black-speckled handlebars. I check the road: empty. And I am off.

Quickly, I'm barreling down Thomas Haynes Drive, past the Ecological Reserve and an indifferent herd of 15 or so cattle. I continue. It's 11 AM and the sun is nearly straight overhead, but a gentle breeze is carrying me — cooling my already-flushed cheeks, combing my loosely-tied hair and peeling the fatigue from my frame and my face, replacing it with calmness. Joy.

I press on, and pedal up the hills. Shoulder-high corn fields pass me on the right. I can see they're nearly ready for picking. The Dover Creek Farm disappears behind me, on my left. Cracks, creases and patchwork cement flow beneath my sneakers as they pedal wildly. And I am free.

This ride feels like living. Like life after numb. The remembering, the carelessness of childhood which is, in its essence, the most true living of all. It's the perfect embrace of beauty. Of time and place. The unhurried presentness a seven-year-old has mastered. She hasn't had time to numb. She hasn't yet descended into the torturous loss of perfect love. She hasn't said goodbye to daddy, mommy. She hasn't yet locked up the first, middle or last parts of her heart. Her eyes are still fierce with life, clear as an untouched glacial spring. She is new. She is here. She is now.

I bend low, careening down a steep hill, when, suddenly, a deer appears in the clearing. As I slow, my foot grazes the spokes, startling the animal who turns and darts from the shoulder just as I pass.

I am well over half-way. My destination: the Junction Café in the heart of town, which later reminds me of the Whistlestop from the film *Fried Green Tomatoes,* which I love.

I am coasting now. I close my eyes, just for a moment. I want to feel the ride without seeing it. Scents and sounds emerge: the soft whistling of wind streaming past my ears, and the smells drawn through my nostrils — a mixture of dried straw, distant manure and the freshness of this morning's early dew.

I reemerge to the startle of a sprinkler throwing a refreshing haze onto my course. It lasts for: *one-mississippi, two-mississippi, three . . . gone.* My legs are beginning to tire, but yesterday's drive reminds me there are only a few miles of straight road ahead. I sigh and reach for my water bottle.

I can feel the greyness fleeing. Colors are becoming more vivid. The greens are a rainbow now: autumn winter tones, lemonade, ginger, palm — the world is spilling over. I can feel my breath slow. Deeper now, deeper. I am slipping, now, along the road, effortlessly.

And later, in the heart of this small northwestern town, I pick a book off the coffee shop shelf and read this:

> "For man, the vast marvel is to be alive. For man, as for flower and beast and bird, the supreme triumph is to be most vividly, most perfectly alive. . . . We ought to dance with rapture that we should be alive and in the flesh, and part of the living, incarnate cosmos."
>
> — D. H. Lawrence, *Apocalypse*

Yes, indeed.

Seeing Life

> "To see life; to see the world; to eyewitness great events; to watch the faces of the poor and the gestures of the proud; to see strange things — machines, armies, multitudes, shadows in the jungle and on the moon; to see man's work — his paintings, towers, and discoveries; to see things thousands of miles away, things hidden behind walls and within rooms, things dangerous to come to; to see women that men love and many children; to see and take pleasure in seeing; to see and be amazed; to see and be instructed."
>
> — Henry Luce

For nearly a century, these words guided *LIFE,* the world's preeminent photojournalism magazine. The motto, written by *LIFE's* founding editor Henry Luce, still resonates.

Poverty. Prosperity. Success. Sickness. Life. It's not tidy. In the flood of birth, the stillness of death, and everywhere in between, life is a mess. People are not strong; we are needy, frail. Alone we falter, together we rise. When we peel back the pages of time, we see this has always been the way. Some things never change.

The Design of People

We are living in an upside-down world. Our wild, strong bodies are sedentary. Intimacy is down, and passivity and isolation are up. Our weak, needy hearts are lonely. We are anxious, tired, overwhelmed, and addicted to technologies we don't fully understand, yet our culture has little to say about something we all feel. We are barreling down the highway of technological "progress," and no one's got the playbook; the rules of the road are yet to be written.

One powerful factor that shields technology from serious examination, says philosopher Albert Borgmann, is liberal democratic individualism: the notion that the individual is to be the judge of what is the good life for him or her. But,

> "Sit as little as possible; credit no thought not born in the open and while moving freely about — in which the muscles do not hold festival."
>
> — Friedrich Nietzsche

perhaps, there *is* a universal good life. And to find it, we must take notes from the signposts all around us.

The Greatest Tablet in the World

The human eye, with its retina and ocular nerves, is magnificent in its design, and, astoundingly, no different today than it was thousands of years ago.

For millennia, human infants have preferred to look at faces that engage them with a mutual gaze; this leads to the formation of the synapses necessary for socializing and founding relationships. It is within our first six months of life that, by probing people's eyes and faces for positive or negative mood signs, we learn who to trust, how to smile, and that eyes are the indicators of where our attention lies.

We are face readers from infancy.

It is said that the eye is the gateway to the soul. Eyes seek out other eyes and, through a series of rapid, repeated scans of the face and body, we decipher the social and emotional information needed to nurture, flirt, sell, teach, converse and make love.

Connected to the eyes is the face, vital to the expression of emotion among humans and among numerous other species. Even more than our eyes, the face is crucial for human identity; any damage such as scarring or developmental deformities has effects stretching beyond those of solely physical inconvenience.

Facial expression is vital for human recognition and communication, and the more than 40 facial muscles in a human face control our expressions as we respond to touch, temperature, smell, taste, sounds and visual stimuli.

Human communication, according to South African academic Rembrandt Klopper, is underpinned by a social survival imperative, because we are not, as he puts it, "merely brains in nutrient-rich vats that can exist in isolation of other humans." We are brains, clad in bodies that interact with other body-clad brains in order to exist, survive and thrive.

Being able to read emotion in another's face is the fundamental basis for empathy. The ability to interpret a person's reactions and predict the probability of others' behaviors is called *mirroring*. The process of learning and teaching, of communication, involves exchange, back and forth. It's why calling someone and getting their voicemail or posting updates online without an exchange of ideas can feel flat.

Reading people is a learned skill. A recent study looking at individuals judging forced and genuine smiles found that older adult participants outperformed young adults in distinguishing between posed and spontaneous smiles. This suggests that with experience and age we become more accurate at perceiving true emotions.

Some studies show that eye contact has a positive impact on the retention and recall of information and may promote more efficient learning. This is important to note because, as we will learn in Chapter 12, people — adolescents in particular — who communicate primarily by text, are losing their ability to read faces and, subsequently, to empathize. (For more on this, read about award-winning actor Astrid van Wieren's work coaching kids and teens in reading facial expression.)

Finally, the human body. According to Klopper, the psyches and metabolisms of modern humans were forged over eons of hunter-gatherer nomadic existence, and our ancient nomadic souls are incongruent with our present-day sedentary existence. It is widely accepted that the average adult needs at least an hour of physical work a day and eight to nine hours of sleep every night to function *optimally.*

Eyes, face, mouth, body. You could say that the human is the original mobile communications device.

The Shape of Things

Our world began with wide open spaces, with room and time for exploration. Imagine, from where you sit: green forest and land expanding as far as the eye can see. In the morning, you wake as the sun rises, and go to sleep as the moon begins its slow ascent. Babies are birthed, food is gathered. Days are filled tending to children, teaching, hunting, gathering, talking, cooking and eating. Ideas are had, land is cultivated, homesteads are raised. The seasons and sun set the course of time. A time to sow and a time to reap. A time to rise and a time to sleep. Then, like now, a time to be born and a time to die. A time for joy and a time for sorrow.

It's said that there is nothing new under the sun. But there *is* something new: the way we *think*. What we hold to be true. We believe as a culture, way deep in our core, that we *are* something new. That *we* are different: more evolved, more advanced, smarter. But are we so different?

The Ties That Bind

We haven't found a way around the nine-month human gestation. There are few things that everyone can agree on, but the fact of our birth is one. The design of procreation requires serious consideration; it is perhaps the only core marker that we all can agree on in terms of a base-level design. You may or may not believe in a conscious creator, God, but you do accept the irrefutable fact that you were born. We were all conceived and born in weakness. While nation, religion and politics may divide, this single fact binds.

We are vulnerable beings. We were born with absolute need. We were cared for and raised, for better or worse, in community and defined in every meaningful way by that experience. The values we hold, the habits we hope to unlearn — these were all learned through relationship.

All seven billion of us share the same biological needs. We each have only one body and only 24 hours in each day. We are constrained by the amount of time we have and by our physical makeup, providing a tangible framework for our lives. We have to eat. We have to sleep. We have to talk to people to work, to play, to procreate, and simply to keep us out of the loony bin. We live in light of this design. It should tell us much about our purpose.

We can view this humanness, and our ensuing mortality, in one of two ways. First, as a limitation, with the connected feeling of being trapped. Or, as a framework, leading to a feeling of freedom within created design. We can try skirting around our design, like sleep hacking and running at treadmill desks, or we can consider them as signposts for how to construct our lives.

> We have real bodies with real limitations in real time. Have you ever considered why?

What People Are For

"To be human," says Jean Vanier, founder of L'Arche, an international network of communities for people with intellectual disabilities, "means to remain connected to our humanness and to reality."

If we are to take human design as the blueprint for our purposes, then we might draw the conclusion that people are *for* three things:

- *We were made to feel.*
- *We were made for relationship.*
- *We were made for work.*

The Fidelity of Work

In Montana a hundred years ago, as Albert Borgmann recounts in his 1984 book *Technology and the Character of Contemporary Life,* warmth in the home came by way of the fireplace, better known then as the hearth.

The heat was not instantaneous because work first had to be done: trees felled; wood sawed, split and stacked; kindling carried and a fire built in the stove or fireplace. As Borgmann points out, it was not an entirely safe method of heating though, because one could get burned or set the house on fire. It was also not easy work; skills and attention were constantly required to build and sustain the fire.

> "A stove used to furnish more than mere warmth. It was a *focus,* a hearth, a place that gathered the work and leisure of a family and gave the house a center. Its coldness marked the morning, and the spreading of its warmth, the beginning of the day. It assigned to the different family members tasks that defined their place in the household. The mother built the fire, the children kept the firebox filled, and the father cut the firewood. It provided for the entire family a regular and bodily engagement with the rhythm of the seasons that was woven together of the threat of cold and the solace of warmth, the smell of wood smoke, the exertion of sawing and of carrying, the teaching of skills, and the fidelity of daily tasks."

In the 21st century, we are taught to think of the elements of the fire and its related work as burdensome, and they are and were undoubtedly so experienced. But the familial work in creating the warmth for the home required skill, attention and strength — there was a fidelity to the work as each family member fulfilled their role, and this gave the work meaning.

Today, the central heating systems omnipresent in modern city life make no demands on our skill, attention or strength, and their presence is largely concealed. A burden has been lifted, or has it?

Unsurprisingly, one of the families described at the beginning of the Introduction was Amish. More than a century ago, their forebears decided to slow their adoption of new technologies. It was an extremely deliberate choice, and their commitment to that choice has been absolute. The city family, on the other hand, did not choose to be without lights and appliances. Instead, against their will, they experienced an unexpected halt to their modern comforts, and in that lack they found something unexpected: joy.

This is something for us to consider.

> "With our time and presence we give love. Simple."
> — Kim John Payne

When the city family found themselves without power, they found themselves with new constraints, and within those constraints they found new opportunities for action and connection. When the power returned, they had the opportunity to choose which appliances and gadgets to reconnect and which to leave behind.

We have the same opportunity.

Good Burdens

What happens when technology moves beyond lifting genuine burdens and starts freeing us from burdens that we should not want to be rid of? We will spend the rest of this book considering this linchpin question, posed by Albert Borgmann.

If we believe that we, as humans, were created for relationship and meaningful work, work that provides for families and serves neighbors, work that engages our bodies and creative faculties, then it follows that we would value a certain kind of burden. Let's call them *good burdens*: responsibilities that tether us to people and the physical world.

"Consider, for instance, the burden of preparing a meal and getting everyone to show up at the table and sit down," says Borgmann, a philosopher at the University of Montana. "Or the burden of reading poetry to one another or going for a walk after dinner. Or the burden of letter-writing — gathering our thoughts, setting them down in a way that will be remembered and cherished and perhaps passed on to our grandchildren. These are the activities that have been obliterated by the readily available entertainment offered by TV" — and every other screen in the 21st-century home.

"The burdensome part of these activities is actually just the task of getting across a threshold of effort," Borgmann continues. "As soon as you have crossed the threshold, the burden disappears."

Have we, by outsourcing the work of our lives, outsourced the living of life?

What if, by outsourcing the work, we are outsourcing the living of life?

The Quotidian Mysteries

Kathleen Norris, the international bestselling author of *Dakota: A Spiritual Biography* and *The Cloister Walk,* echoes Borgmann's sentiments and turned heads with her Madeleva Lecture: "The Quotidian Mysteries: Laundry, Liturgy and 'Women's Work.'"

Before an audience of academic women, she argued: "It is precisely these thankless, boring, repetitive tasks that are hardest for the workaholic or utilitarian mind to appreciate." Children, on the other hand, approach the washing "for the sheer joy of it — the tickle of the water on the skin... It is difficult for adults to be so at play with daily tasks in the world."

As Norris paints so poetically, joy can be found in repetitive physical work because our minds are able to wander — and we are reminded of our smallness in the gravity of the mundane.

"Repetition is reality," wrote Søren Kierkegaard.

And, indeed, it is. We might enjoy a flight of fancy or experience a breathtaking vacation but, in the end, we return to dishes. In Norris's view, this isn't to be shunned.

Sylvia Plath, in her semi-autobiographical novel, *The Bell Jar,* wrote that her character, Esther Greenwood, wanted to be done with things, at once, forever. She wanted to be free of reality. And tragically, in the end, she was.

As I write this book, I have three children, ages four, two, and six months. Every day, their all-so-tangible needs clamor against me, stirring me to rise again and again to the occasion. I am no stranger to the discontentment this kind of thankless work can breed, but in it I have found profound moments of gratefulness. My children keep me alert, attached to the real world. Without them I would spend my days the way I tend to spend my evenings: sedentary, eyes glued to screen. Instead, I am forced *awake.*

What if we learned to be at play in the tasks of the day?.

Sedentary Bodies, Sedentary Minds

When a body is not used, what happens? It seems that every day there is a new study or article outlining the pitfalls of a sedentary lifestyle. We sleep, we drive, we sit. When a body is not used, such as in the case of a patient relegated to a hospital bed, it literally begins to decompose. Many of us have witnessed this in aging parents or hospitalized friends. I've seen it with my own eyes as my mother's body diminished, losing the ability to stand or sit up, under the burden of colon cancer.

Put simply, our bodies were made for moving. But, instead, many of us are sitting 15 hours a day, spending the majority of our waking moments on the couch, in an office chair, or in the car. By not moving, our risk of heart disease has increased by up to 64 percent; we've shaved off seven years of quality life, and we're more at risk for certain types of cancer.

Thanks to Uber, Lyft or Sidecar, we can call a car to get us with the tap of an app. We no longer need to think about the kinds of clothes we like to wear; an algorithm can do that. Clothes dirty? There's a pick-up and delivery service for that. Hungry? Order takeout. Sick of dishes? Stock up on paper plates. I heard of one guy who didn't have to leave his apartment for an entire year by availing himself of the wonders of eBay and FedEx.com.

Next stop: the "smart house."

"Sophistication reaches a critical mass as it does in a so-called smart house," says Albert Borgmann in an interview with David Wood at the University of Chicago (1999) "where every last and least domestic chore and burden is anticipated and taken over by an automatic device, inhabitants become the passive content of their sophisticated container. The vision of such an environment often carries the implied promise that people will use their disburdened condition creatively and inventively. But assuming that in the smart house the blandishments of cyberspace will present themselves with even greater diversity and glamour, most people

We know what happens to a muscle when it is not used: it atrophies and dies. Our brain is our greatest muscle.

will likely do what they now do — they will immerse themselves in the warm bath of electronic entertainment."

And so we do.

We understand the causes and effects of the sedentary body, but, as the Internet takes over and our memory is needed less and less, what happens to our sedentary *minds*?

> In the absence of physical work, we turn to consumerism. Remember the Capitol in the *The Hunger Games*?

I share Albert Borgmann's view that humans are essentially embodied beings and therefore cannot escape their bodies no matter how or what they think of themselves.

"The crucial error these days is not to think of oneself as a machine but to shift one's moral center of gravity into a machine of sorts — cyberspace," he says. And the temptation to entrust our curiosity and desires primarily to the Internet will only grow. "To do so is not to commit a cognitive error but to become an accomplice in the diminishment of one's person and one's world. Just as you cannot escape your body, you cannot really and finally escape reality." [And, I argue, nor should you want to.] "But you can degrade to utilities what should be celebrated as the splendor of tangible presence."

As we slip further and further away from the fidelity of our design, the world has become complicated, and we have grown aimless and weary in the process. But we don't have to remain here; we can untangle the web, pull back the layers of our complicated 21st-century lives and chart a new course.

> "Whether one life is enough or not enough, one life is all we get, at least only one life here, only one life in this gorgeous hair-raising world, only one life with the range of possibilities for doing and being that are open to us now ... we tend to live as though our lives would go on forever. We spend our lives like drunken sailors."
>
> — Frederick Buechner

The Law of Unintended Consequences

Black Mirror, the Emmy award-winning BBC series written by Charlie Brooker, satirizes real-life cutting-edge technologies. "Over the last ten years, technology has transformed almost every aspect of our lives before we've had time to stop and question it," say the show's creators. "In every home; on every desk; in every palm — a plasma screen; a monitor; a smartphone — a black mirror of our 21st-century existence. Our grip on reality is shifting."

In the episode, "Be Right Back," Martha is devastated when her partner, Ash, is killed in a road accident the day after they move out to a country cottage. At his funeral, Martha's friend Sarah tells her about a new service that allows people to stay in touch with the deceased. By using all his past online communications and social media profiles, a new "Ash" can be created — as a voice on her phone and finally, in the most-advanced version, as a bot. Martha is disgusted by the concept but then, in a confused and lonely state after learning she is pregnant, she decides to talk to "him."

When we draw our technological future out to its logical conclusions, as Charlie Brooker has done in *Black Mirror,* we see a future not far out of reach. The technology to create a bot in our dead spouse's image to help us grieve is not an impossibility.

But it is also not an *inevitability.*

The Future Is Not an Inevitability

Many people in Silicon Valley speak about our technological future like this as an inevitability. In her book *Dot Complicated,* Randi Zuckerberg (sister to *that* Zuckerberg) writes about pursuing an online life as fulfilling as our offline one. But, in light of our physical bodies and deeper humanity, how could that ever be true and why are we working so hard to make it so?

As my mother has always reminded me, life is choices. We, through our acts of creation, condoning and consumption, are choosing our collective future.

The technologies that save lives, the databases that expedite hospital work, the search and rescue capabilities enabled by global positioning systems — these are all tools that *help,* that bless and serve our lives; they are about betterment, they lift genuine burdens. These same technologies, however, are being adapted and repackaged in myriad ways that are drawing us further away from one another, from meaningful engagement with the world and, ultimately, away from what it means to be human.

But all of this comes down to belief. If we believe that people are here to consume for consumption's sake, that we are all a string of random molecules smashed together for some unknown future, then there is little binding us to moral and social implications. But if we believe that there is some design to our world, to our bodies, minds and inner lives, then we must pause and consider our future in light of this question: *What are people for?*

Instead of altering or avoiding these purposes, the relationships and work that gives us meaning, we should be spending our lives seeking to live more deeply from them.

Why did Segway (the two-wheeled, self-balancing personal transport system) fail as a globally embraced human transport system? Because we were made to *walk*.

2

Information Overload

How We Got Here

In the span of 150 years, we have moved from homing pigeons to the telephone to featherweight wearable computers. Electrical and electronic technologies have altered the nature of communication and the shape of our lives. It is impossible to overemphasize the systemic shifts we have undergone in only a few generations.

The history of communication stretches from prehistoric forms of art and writing to satellites and the Internet. Mass communication began when humans could transmit messages from a *single source* to *multiple receivers.*

According to Marshall McLuhan, communication technologies have always been the primary cause of social change. McLuhan, a Canadian philosopher of communication theory, is known for coining the expressions "the medium is the message" and "the global village," and for predicting the World Wide Web almost 30 years before it was invented. According to him, the media of the epoch defines the essence of the society, and he divides history into four periods: the Tribal Era, Literate Era, Print Era and Electronic Era.

Dr. Read Schuchardt explains these four stages as taking us from (1) speech as our primary medium of communication from birth to (2) frozen speech in writing to (3) frozen, multiplied speech in printing to (4) frozen, multiplied and distributed speech in electronic form. We have rapidly progressed from electricity and the light bulb to the telegraph and to our now-ubiquitous handheld instantaneous communication devices.

As industrial values have gained dominance over time, social and psychological problems have risen with them. Industrialization serves the needs of industry, not humans. Our dominant capitalist career culture does not pay enough attention to enduring human needs for love, friendship, leisure, community life, rest, even delight.

Let's consider our trajectory.

The timeline that follows is adapted from a series of talks given by Dr. Read Schuchardt: "Living in a World with No Off Switch."

Tribal Era

Ravi is sitting cross-legged, talking to his sister.

The imperfection of speech, which stimulated inventions and allowed for a simple transfer of ideas, eventually led to new modes of communication which increased both the range at which people could disseminate ideas and also the longevity of the information. All of those inventions were based on the *symbol*: a conventional representation of a concept.

The earliest telecommunication — the transmission of signals over a distance for the purpose of communication — began thousands of years ago with the use of smoke signals and drums throughout Africa, America and parts of Asia.

From time to time over the past 200,000 years, new forms of communication have emerged for survival's sake to help humans cope with greater cultural complexity. Nonverbal communication (popularly known as body language) emerged first and was complemented by verbal communication during humankind's hunter-gatherer existence. About 10,000 years ago, our nomadic ancestors learned to domesticate plants and animals in the Northern Mediterranean climatic zone. After that, ancient forms of graphical representation (some of which can still be seen in cave art) successively developed into pictographic and alphabetic writing systems.

Literate Era

Blaise is writing a letter.

The oldest-known forms of writing were based on pictographic and ideographic elements. The invention of the first writing systems are attributed to the Bronze Age in the late Neolithic period around 4,000 B.C. The first writing system is generally believed to have been invented in pre-historic Sumer (modern-day Iraq and Kuwait) and developed by the late 3,000s B.C.

5th Century B.C.

Harold Innis, a political economist and pioneer in communication studies, argues that Greek civilization flourished due to a balance of oral and written forms of communication. Much of the world remained illiterate except for the upper classes.

Pigeon Post

As a method of communication, using pigeons as messengers began with the Persians who are credited with the art of training birds. The ancient Greeks conveyed the names of the victors at the Olympic Games to various cities through pigeon post, and the Romans used pigeons to aid in their military quests. In fact, Frontinus wrote that Julius Caesar used pigeons as messengers in his conquest of Gaul.

Print Era

Betsy is reading the newspaper.

1045

The world's first movable-type printing technology was invented and developed in China between the years 1041 and 1048. In Korea, the movable metal-type printing technique was invented in the early 13th century during the Goryeo Dynasty.

Clocks were invented in 12th century. For the first time, people understood that the day is broken up into hours, minutes and seconds. This is a pivotal philosophical idea; time had become a kind of abstraction. It's the beginning of the *interchangeable part,* wherein we divide things into discreet, repeatable segments. From the clock, says Dr. Schuchardt, it's a short hop to the printing press, in which block letters and hot metal press are developed — the first interchangeable parts in physical form.

1450

In the West, the invention of an improved movable-type mechanical printing technology is credited to the German printer Johannes Gutenberg in

1450. The printing press displaced earlier methods of printing and from there began the mass production of books, and the mass distribution of ideas. A single Renaissance printing press could produce 3,600 pages per workday, compared to just a few copying by hand.

The book is the first mass-produced artifact in cultural history in which a repeated number of physical objects is produced. Rapid production stems from here.

1760–1840

The Industrial Revolution marked the transition to new manufacturing processes. Including the shift from hand production to machines, the introduction of chemical manufacturing and iron production processes, and the increasing use of steam power and coal. The average income and population began unprecedented sustained growth.

During the era of the Industrial Revolution, written communication was optimized and stretched to its very limits to cope with accounts of new geographical, technological and scientific discoveries. Free public education was established, and the emergent society tended to create smaller family units. Key functions of the family, such as education and care of the aged, were parceled out to new specialized institutions, in order to free workers for factory labor.

> "Above all, the new society required mobility. It needed workers who would follow jobs from place to place . . . Torn apart by migration to the cities, battered by economic storms, families stripped themselves of unwanted relatives, grew smaller, more mobile, more suited to the needs of the [workplace]."
>
> — Alvin Toffler

American writer Trudelle Thomas, author of *Spirituality in the Mother Zone,* identifies a helpful line of thought in sociologist Arlie Hochschild's book, *The Second Shift,* that explains key shifts in this era:

> "For millennia, humans lived according to the sun and the rhythms of the human body with its need for both rest and

> busyness. Only with the rise of industrialization did great masses of people live by the clock, as more and more men took jobs in factories and offices . . . In the career culture, time is measured and standardized and seems to move at a faster beat. [It] is ruled by the clock, with time demarcated into schedules meant to measure productivity and commitment . . . Productivity, efficiency, status, material goods, and standardization are prized values. The home culture, like preindustrial societies, is by necessity based on seasonal and biological cycles."

As Thomas points out in her book, a division arose where career culture became more prized than the home culture, where the caring professions are void of status, money and the industrial understanding of opportunity. The result has been the devaluing of family life and the imposition of career values on home life.

Productivity, efficiency, status, material goods and standardization rule the day whether we are preschoolers, CEOs, students, or stay-at-home parents. In Thomas' view, this one-size-fits-all pace doesn't compute, and our mental and physical health have suffered as a result.

1839

Electricity and lights allow us to work anytime, every day.

1840

First census in the United States of America. On the census form, there is a single category for mental health problems: idiocy/mania.

Electronic Era

Maria is running to answer the phone in the hall.

1844

Telegraph. For the first time in the history of humanity, the speed of communication is no longer limited by the speed of transportation. Now, all of a sudden, content is unhinged from the form: a message can travel faster than a person can.

1851

Sending messages by pigeon post had a considerable vogue amongst stockbrokers and financiers. In 1851 London, German-born Paul Julius Reuter opened a business transmitting stock market quotations between London and Paris through the new Calais-to-Dover telegraph cable. Reuter had previously used pigeons to fly stock prices, and continued to do so between Aachen and Brussels, until a gap in the telegraph link was closed.

1850s

All but the largest metropolitan daily newspapers were just four pages and featured very little illustration. Newspapers followed the format of the book and were intended to be read comfortably by readers who, the publishers presumed, would read every word. Because papers were only four pages long, this expectation was reasonable. Readers were assumed to be rational, active citizens who would be persuaded by facts and strong arguments and would be discerning about falsity and fallacy.

> The media, first through the printerly form of the newspaper, transformed the average citizen from a partisan voter to an industrial consumer and, finally, to a modern spectator.

But then, newspapers shifted from being a type of courtroom filled with persuasion and thoughtful argument to a *marketplace.* Newspapers began to stock themselves like aggressive shops, thinking of their front pages like shop windows. The intent to sell began to win out over the intent to inform and educate.

1861

A German named Johann Philipp Reis manages to transfer the phrase "The horse does not eat cucumber salad" electrically over a distance of 340 feet with his Reis telephone.

1876

Alexander Graham Bell makes the world's first long-distance telephone call — about six miles — between Brantford and Paris, Ontario, Canada.

1880

Telephone service is initiated; and thus the idea of assigning a number to a specific human being is born.

The second census is taken, and there are seven categories for mental disorders: dementia, dipsomania (uncontrollable craving for alcohol), epilepsy, mania, melancholia, monomania (partial insanity), and paresis (a disorder affecting the brain and central nervous).

1882

> "God has died and we have killed him."
>
> — Friedrich Nietzsche

1885

Building on 200 years of invention, photographic film was pioneered by American George Eastman. Eastman began manufacturing paper film in 1885 before transitioning to celluloid in 1889. His first camera, the "Kodak," hit the market in 1888. It was a simple box camera with a fixed-focus lens and single shutter speed. Affordably priced, the Kodak appealed to the average consumer who would shoot 100 exposures before sending the camera back to the factory for processing and reloading. Photography forever changed fixed print — newspapers, magazines and books — by offering immediacy, realism and emotion.

1900s

Mass expansion of electricity use in early 1900s.
The census lists 22 mental health diagnoses.

The Amish

Whereas in contemporary society the default is set to say "yes" to new things, Old Order Amish societies default to "no," governed by the primary aim to strengthen their communities. When cars first appeared at the turn of last century, the Amish noticed that drivers would leave the community to go shopping or sight-seeing in other towns instead of shopping locally and visiting friends, family or the sick on Sundays. Therefore, the ban on unbridled mobility was aimed to make it hard to travel far, and to keep energy focused in the local community. (Read more about the Amish in Chapter 14: Here on In)

1914

First North American transcontinental telephone call.

1918

Vladimir Lenin sees a connection between electricity and mental health issues. The effects of electricity and industrialization and their impact on humans begin to be questioned, as, for example when Charlie Chaplin is turned into a "machine man" in the film *Modern Times.*

1926

First transatlantic telephone call, from London to New York.

1927

World's first videophone call via an electro-mechanical AT&T unit, from Washington, D.C. to New York City, by then-Commerce Secretary Herbert Hoover.

1927

Television

1948

Polaroid Model 95, the world's first viable instant-picture camera — precursor to digital cameras and Snapchat, with its online disappearing pictures.

1950s

In the late 1950s, early networks of communicating computers included the military radar system Semi-Automatic Ground Environment (SAGE).

One hundred and six mental health diagnoses are listed in the newly published *Diagnostic and Statistical Manual of Mental Disorders* (*DSM*) published by the American Psychological Association.

1952

The term "anti-depressant" is coined, and pharmaceutical companies determine that 50–100 million people are suffering from depression and can be helped with prescription drugs.

1968

The new edition of the *DSM* lists 182 mental disorders.

1969

The first-ever computer-to-computer link was established on ARPANET (Advanced Research Projects Agency Network), the precursor to the Internet, on October 29, 1969. ARPANET was to be used for projects at universities and research laboratories in the US. The program is now called DARPA (Defense Advanced Research Projects Agency), part of the US Department of Defense.

1971

The first network email was sent by US programmer Ray Tomlinson to himself. It was the first system able to send mail between users on different hosts connected to the ARPANET. (Previously, mail could be sent only to others who used the same computer.) To achieve this, he used the @ sign to separate the user's name from the machine's, and it has been used in email addresses ever since.

1973

By 1973, email constituted 75 percent of ARPANET traffic.
First modern-era mobile (cellular) phone.

1978

Transportable telephones, known as "luggables" (because they were so large), were essentially car phones, with a handset, antenna and electrical components packaged together in a carrying case. The sets were adopted early by news organizations to send direct reports to the studio from the field. The next generation of the cellular phone was lovingly referred to as "the brick."

> "There is no reason anyone would want a computer in their home."
>
> — Ken Olsen, founder of Digital Equipment Corporation, 1977

1980

The beginning of the computing revolution.

Steve Jobs, in an interview with *Rolling Stone:* "Well, it's like computers and society are out on a first date in this decade, and for some crazy reason we're just in the right place at the right time to make that romance blossom [with the Macintosh computer]. We can make them great, we can make a great product that people can easily use."

The *DSM* lists 265 mental disorders.

1982

SMTP email, computers and networks emerge.

> "In many ways the spread of electronic transmission of bits of information marks the end of the simple notion of communication as the transfer of a message from a sender to a receiver."
>
> — Ursula Franklin

1983

The Internet. The omnipresent decentralized global system of interconnected computer networks was originally designed to fulfill the military's need for the instant and resilient exchange of data.

Ursula Franklin explains it this way: "It is essential to remember that computers not only compute, manipulate numbers, and store data for fast retrieval, but they also transform information — words, sounds, and images — into bits that can be transmitted in digitized form and be received and recombined into words, sounds, or images at the point of request." Experts believe that, in many ways, this was like inventing writing again.

It is a virtual (nonhuman and unreal) environment for the transfer, storage and retrieval of messages.

> "It is important to view the Internet as a continuation of preceding technological developments that have tried to change the constraints that time and space put on human pursuits."
>
> — Dr. Read Schuchardt

1992

> "This species has amused itself to death."
>
> — From the song Amused to Death. by Roger Waters

People begin to get very disappointed with the state of affairs.

1993

Early smartphones, devices that combined telephony and computing, were brought to market. The term "smartphone" first appeared in 1997, when Ericsson described its GS 88 "Penelope" concept as a Smart Phone. The original features included a personal digital assistant (PDA), a media player, a digital camera and GPS navigation unit.

1994

The *DSM,* with 297 mental disorders, is now nearly 900 pages long.

1998

Mobile satellite hand-held phones and the proliferation of the Internet. Also, the year I got my first email account.

2007

Apple Inc. introduced the iPhone, one of the first mobile phones to use a multi-touch interface, notable for its direct finger input instead of the stylus or keyboard synonymous with other smartphones at the time. Smartphones were released to market with all of the features of a laptop: Web browsing, Wi-Fi, plus third-party apps and accessories.

Five percent of men (15 million) and 15 percent of women (33 million) are taking antidepressants in America.

2011

We expect to be online all of the time, and we expect to be reachable everywhere. The average cable television provider offers 500 channels.

2012

There are more pieces of digital content in the world than there are grains of sand on every beach in the world. (Source: "Digital Universe" report conducted by analyst firm IDC, 2012.)

The multiplication and fragmentation of facts continues at breakneck speed. This is reflected in some unexpected ways. In Dr. Read Schuchardt's view, it "is revealed in the rise of mental health diagnosis and the fragmentation of the Christian church from two recognized denominations to more than 23,000."

2013

The 5th edition of the *DSM.* Revisions since its first publication in 1952 have incrementally added to the total number of mental disorders recognized, while removing those no longer considered to be mental disorders.

In a *New York Times* editorial, psychiatrist Allen Frances criticized the new *DSM* for medicalizing normality, resulting in "a glut of unnecessary and harmful drug prescription." Some of the changes that concerned him include listing the following as "disorders":

> Disruptive Mood Dysregulation Disorder — for temper tantrums; Major Depressive Disorder (includes normal grief); Minor Neurocognitive Disorder — for normal forgetting in old age; Adult Attention Deficit Disorder (encouraging psychiatric prescriptions of stimulants); Binge Eating Disorder — for excessive eating; and Generalized Anxiety Disorder (includes everyday worries); Behavioral Addictions — making a "mental disorder" of everything we like to do a lot.

One billion smartphones in use worldwide. Ninety percent of people keep their mobile phones within three feet of them 24 hours a day.

Twenty-eight percent of Apple product users and 23 percent of Android users would rather go without seeing their significant others for a week than part with their phones, according to a 2012 TeleNav survey. A Yahoo survey reveals that nearly 15 percent would rather give up sex entirely than go even a weekend without their iPhones.

2014

Facebook turns ten. Every day 2.5 quintillion bits of information are added to the Internet.

2015 and beyond

- *Remee*: the sleep mask that lets you control your dreams.
- *Stinky Footboard*: for playing videogames with your feet.
- *Skycube*: Launch your own tiny, tweeting satellite into space.
- *Project Hexapod*: A giant rideable walking robot.
- *Puzzlebox Orbit*: A brain-controlled helicopter.

- *Grief robots.*
- *Immersion virtual reality.*
- *Iron man suits custom-made for US Army.*
- *One-year-olds with personal computer tablets.*

(I am not making this stuff up.)

We were born to the earth, but we want to fly to the stars. And now, online, we can prebook a ride to the moon for a cool $100 mil.

3

Better Off?

Yes, BUT

My focus is less on setting limits than it is on creating the positive conditions in which technology becomes less compelling and different kinds of engagements thrive and flourish.

— Albert Borgmann

SO, WHERE ARE WE?

Today, we can talk with friends on the other side of the world in real time. Parents can connect with their children via Skype or FaceTime while traveling for work. Surgeons can save lives with the most advanced medical technologies. We can send and receive important files in milliseconds, allowing flexibility in our work. Police can capture wrongdoers with the help of viral video.

In many ways our lives are helped, our worlds expanded.

At the same time, we don't have to leave our homes anymore; we can order all we need with the tap of an app. We can work 24 hours a day, anytime, anyplace. Tailor-made advertisements prompt us to consume more than ever. Kids are demanding iPhones to bide their time in strollers, restaurants and cars rather than engaging in conversation or simply allowing their minds to wander. Our email boxes and Twitter feeds act like cocaine, setting our bodies aflutter with measurable levels of anxiety.

In other ways, our lives are hurt, our worlds are shrinking. The truth is, the Internet is a mixed bag.

What Has the Internet Given Me?

The primary thing the Internet has given me is a means to consume more information, more products and more entertainment — faster.

The secondary thing it has given me is an ability to work from anywhere. This is something I value greatly. I can research and write from home or a nearby coffee shop, and still spend the majority of my time raising my children. This is no small thing. I have access to research, the ability to contact faraway interviewees free of charge thanks to Skype and other chat services, and I can file stories in nanoseconds via email. With the help of Siri, I can even pitch stories while nursing a baby. *Been there, done that.*

The final thing the Internet has given me is a way to connect with people far away, family and strangers alike. In my experience, these means of connecting can be quite limited but, living away from the eleven members of my immediate family, valuable nonetheless.

There are also all of the everyday expediencies, like the Web-based databases that ensure my sick kid gets in front of an emergency room doctor as fast as possible, for which I am indebted.

Pause for a moment and answer this question as honestly as you can: What has the Internet given me?

Internet: The Good

"The fact that you're able to read this — and I'm able to write it — on a computer with Internet means we've all hit humanity's jackpot in some regard," says Shane Snow, a journalist and CCO of Contently. "We're literate. We live in a time where technology has improved the quality of life dramatically over any other period in history. And we can afford to instantly access the greatest trove of information that ever was."

Barry Wellman and Lee Rainie share Snow's enthusiasm. In their book *Networked,* Wellman, of University of Toronto's NetLab, and Rainie, Director of the Pew Research Center's Internet & American Life Project, reveal how, in their view, "large, loosely knit social circles of networked individuals expand opportunities for learning, problem solving, decision making, and personal interaction.

"The new social operating system of 'networked individualism' liberates us from the restrictions of tightly knit groups," they argue; " it also requires us to develop networking skills and strategies, work on maintaining ties, and balance multiple overlapping networks."

Rainie and Wellman focus on the capacity of the Internet to empower individuals and on how the Web has expanded personal relationships beyond households and neighborhoods, transforming work into more team-driven enterprises through the creating and sharing of content.

Internet as Liberator

Clive Thompson, a columnist for *Wired* and author of *Smarter Than You Think: How Technology is Changing Our Minds for the Better,* makes the claim that the Internet age has helped us learn more and retain more information, producing a radical new style of human intelligence. He writes: "In the early 1990s, I believed that as people migrate online, society's worst urges might be uncorked . . . Certainly some of those predications have come true . . . but . . . I didn't foresee all the good stuff. And what a torrent we have: Wikipedia, a global forest of eloquent bloggers, citizen journalism, political fact-checking."

> "When it comes to technology in the home, more than a third of the surveyed executives view it as an invader, and about a quarter see it as a liberator."
>
> — *Harvard Business Review*, March 2014

> The three major shifts in the Internet age, according to Clive Thompson: Infinite Memory, Dot Connecting, and Explosive Publishing.

In *Smarter Than You Think* Thompson documents how every technological innovation — from the printing press to the telegraph — has provoked the same kind of panic that life will never be the same again, but that as we adapt — as we always have — we learn to use the new and retain what's good of the old.

My own grandfather in London, Ontario (now 80) got a laptop at the ripe old age of 75, and for him there's been no turning back. He told me by email: "I enjoy the Internet immensely as it allows me to be in contact with my children and grandchildren as often as I want — and that is frequently. It also allows me to visit galleries and faraway places. I can be in touch with my friends far and near and have become reacquainted with school friends the world over, and this would not have happened without the Internet."

A consummate volunteer, he now receives his schedule for Museum London and other important notices via email or wiki. All in all, he says it gives him a lot of pleasure that he would not want to do without. With the help of the young man who provides him tech support, my grandfather is fully integrated online, and his life is the better for it.

So is mine.

BEING HEARD: FASTER. BOLDER. LOUDER.

> When I began blogging — nearly ten years ago — there was a lot less noise on the Internet. There was a distinct sense of being heard, even if your readers were few. But, a decade later, this sense has precipitously declined as a billion Twitter handles and Instagram accounts have cropped up. There is too much volume, so much clutter. If you want an audience, you must be BOLDER, you must be FASTER, you must be LOUDER and, above all, you must be CLEVER.

I remember when I started blogging on an old site called xanga. I had journaled all my life but was deeply fearful of other people reading my words. Wanting to pursue a career in journalism, this was a hurdle I had to face. I needed something to get over the hump. So, I signed up for a free blog and wrote my first post: a reflection on my neighborhood and a mediocre poem. And with that, I hit "publish." The only people who knew I was blogging were my family and a few select friends. *This wasn't so bad.*

At the time, I was working on contract as a writer for the CBC, Canada's public broadcaster. I was assigned wickedly fun projects like researching everything you could think of to do with water. Anything. *Water balloons. Walruses. Water polo.* And as I worked I listened to the then-cutting edge Web radio station, CBC Radio 3, produced only a few floors below me in the bowels of the cement block that was CBC Vancouver. It was from there, during my lunch break, that I composed my third blog post:

> Oh my. I am sitting in the CBC sipping a frappuccino and listening to Kid Koala's "Drunk Trumpet." (yes, I'm working.)

> Could life get any sweeter than this? The sun is shining and I have a view of our stunning downtown library. Wheeeeee!!! Sorry — just had to get that out of my system. Have a fabu day.
>
> Listen to www.cbcradio3.com/ to sweeten your day.

Nothing profound, but I had an outlet, an (albeit small) audience, and my confidence was growing. I was finding my voice.

> "Often I need to occupy a safe space. And I have found safe spaces like that, time and time again, on the Internet."
>
> — Esther Emery

Being heard is not insignificant; in fact, it is a cornerstone to nurturing a sense of belonging.

Call to mind your five-year-old self chattering away before your dad, your teacher, or the old lady on the bus. What were you seeking? What happened when their gaze met yours, when they knelt to your level, asked you questions, praised your thoughts? I see it in my daughter: her chest expands, eyes warm, face relaxes, as she realizes: *you see me.*

Nothing says "I love you," says "I belong" more than the moment you read kindness in somebody else's eyes. In his 1943 paper, "A Theory of Human Motivation," Abraham Maslow found that in the absence of this love or belonging element, many people became susceptible to loneliness, social anxiety and clinical depression. A lack of belonging leads to personal and interpersonal discord.

In a recent movement to help promote peace in the Middle East, a number of organizations have established "peace camps" or similar conflict-resolution programs that bring Israelis and Palestinians together to foster greater understanding of the opposing group. Each gathering is an opportunity for members of each group to share stories about their lives with members of the other. A new study from MIT neuroscientists shows that the benefits from this exchange are much greater when members of the less empowered group share their stories with the traditionally dominant group than when the reverse occurs. "If that sense of being neglected and disregarded and taken advantage of is the biggest obstacle to progress, from their perspective, then you can partly address that by providing an experience of being heard," says Rebecca Saxe, an associate professor of brain and cognitive sciences and associate member of the McGovern Institute for Brain Research at MIT.

The finding, published online in the *Journal of Experimental Social Psychology,* supports the idea that the biggest barrier to reconciliation is the disempowered group's belief that their concerns will continue to be ignored. When the less dominant group had the opportunity to exchange ideas and feel meaningfully heard, things began to change.

The Underdog Takes the Stage

The story of the Internet can be told or understood as a David and Goliath kind of tale. Egypt's Arab Spring, Bangladeshis calling out injustices and the courageous Chinese students mounting an online movement that shut down a $1.6 billion toxic copper plant — these are just a few examples of the voice of a few reaching the ears of a thousand.

Closer to home, we see again and again the Internet giving natural-born introverts a safe place to take the podium.

"Social media for introverts," suits Calgary-based publisher Janine Vangool just fine. An illustrator by trade, Vangool launched UPPERCASE, an international magazine for the *creative and curious* in 2009. By all accounts, she's done it all right. Combining well-researched articles, meticulous design and a prolific online presence, Vangool's fledgling Canadian publication has grown to an award-winning global magazine with a cult following. The persona Vangool exudes online is bold. She tweets daily to more than 13,000 "followers" and has used every tool in the online promotion box. But, on the day we meet in person, inside Calgary's Art Central building, she is quieter than expected. But no less powerful.

The Internet, Susan Cain explains in her book *Quiet: The Power of Introverts in a World That Can't Stop Talking,* allows quietly confident introverts to comfortably communicate to thousands. Vangool echoes this sentiment.

All over the Web we find examples of the lonely, the underdog, the minority, the recluse harnessing their powers and spreading messages like wildfire. This would make the earliest Internet pioneers proud.

The Dreamers: Utopian, Democratic, Free

Joseph "Lick" Licklider was an American psychologist and computer scientist and is considered one of the most important figures in computing history. As an Internet pioneer, he saw an interconnected future where computers would empower individuals, instead of forcing them into rigid conformity. He was a visionary in his rare conviction that computers could become "not just superfast calculating machines, but joyful machines: tools that will serve as new media of expression, inspirations to creativity, and gateways to a vast world of online information." (Waldrop)

In 2011 a global network of gamers took a puzzle that had baffled HIV scientists for a decade, made it an online game and solved it collaboratively — in only *one month*. This was the joy Lick was dreaming of. (Read more about collaboration in Chapter 13: Making Space to Create.)

Those on the fringe, many of whom once felt alone, have come together; they are finding belonging online. Ironically, at the same time, many of us have found the reverse, that we are lonelier than ever.

Internet: The Bad

Sherry Turkle was one of the Internet's early champions. In her 1996 TED (Technology, Education and Design) talk, "Celebrating Our Life on the Internet," she was excited, as a psychologist, to take what people were learning in the virtual world and apply it to the physical world. It made the cover of *Wired*. Turkle, who has written extensively on the nature of human relations on the Internet, who evangelized the Internet, who loves getting text messages from her daughter, returned to the TEDTalk stage in 2012 to change her tune from celebratory to chaste. *There may be a problem.*

Alone Together

In her most recent book, *Alone Together: Why We Expect More from Technology and Less from Each Other,* Turkle is still excited about technology, but is extremely concerned that we are letting it "take us places we don't want to go." She's interviewed hundreds of people about their online habits, and she's come to the conclusion that "the little devices in our pockets are so psychologically powerful that they don't even change what we do, they change who we are." Things we do with them used to be odd, like texting in board meetings, classrooms and even funerals, but now they seem normal.

"People want to be with each other, but also elsewhere. People want to control exactly the amount of attention they give others, not too much, not too little." She calls it the Goldilocks effect. But the distances that feel right for some people in some situations can be totally wrong in others.

"There is a feeling that conversations are difficult because we don't have the ability to edit as we talk, and so can't present the exact face that we'd like to, the way we can online," she continues, "Human relationships are rich, and they're messy and they're demanding. And we clean them up with technology."

> We are not only living at arm's length from each other, we are living at arm's length from ourselves.

Stephen Colbert once asked her, "Don't all those little tweets and texts, all these little sips of information, add up to one big gulp of real conversation?" Her answer is no. All of these little bits work very well for many things, but they don't work for the real task of learning deeply about each other or ourselves.

Arm's Length

When it comes to text-based, or any other mediated form of communication, we can choose our response time. An email may be sidelined for weeks, a voicemail returned days later, a text message, whenever. We can avoid incoming messages by leaving them unread, choosing not to answer an incoming call, and ignoring someone's attempt at Facebook chat. The more mediated our communication becomes, the more we are able to live our lives at arm's length — even in our closest relationships.

When my husband was going through a particularly difficult time a few months ago, in a bluster of tears, he reached out to his best friend through their usual medium: BlackBerry Messenger (BBM). As their short messages pinged back and forth, the sense of aloneness ebbed but it didn't subside. My husband needed more, but dialing the phone number was too great a hurdle.

> "These phones in our pockets are changing our minds and hearts," says Susan Turkle, "removing ourselves from our grief and from our reverie."

With all of the modes of communication at our disposal, we tend to choose the path of least resistance: we text.

Last year, my dad got an iPhone. Until that point, he remained one of two people in my life that I communicated with solely, if not in person, by phone. No texts. No Facebook. No emails. Only his and my voice connecting in real time across the wide expanse of Canada. (Thank you, cell phones.)

Now my father's and my connection is like every single other adult relationship in my life: scattered across a variety of platforms — and it doesn't feel any richer. I can see photos of him from a recent fishing trip in the Queen Charlotte Islands, but haven't heard him describe what it was like to be there — the weight of the catch in his arms, the ridiculous story about the boat driver. Similarly, because he's read my well-crafted snippets online, I'm not inclined to recount them. When our voices meet again, we fill in the gaps rather than telling the story.

When our voices meet again, we fill in the gaps rather than telling the story.

Keeping life and others at arm's length, and protecting ourselves in the process, is nothing new. In his short story "A Man in a Shell," Anton Chekhov (1860–1904) wrote: "And you know, the way we live in the city, the closeness, crowded together, how we sign unnecessary documents, play cards, isn't that really a shell? And the way we lead our whole lives among loafers, people pursuing lawsuits fools . . . talking and listening to all manner of nonsense, isn't that really a shell?" (Source: *About Love,* translated by David Helzwig)

Year after year, century after century, our shells are taking new form.

Our Intermediaries

"Etymologically, *media* means something like 'things in the middle' or 'in-between things.' One might look upon the *New York Times* as a paper-and-ink product, or as the text of today's *Times,* or as the New York Times Company. It is, of course, all of these things and more. If we keep its middleness in mind, it is the network of relationships between all of the producers and receivers and buyers and sellers that connect through it." (Nerone and Barnhurst)

We have enabled our online media to sandwich themselves squarely between us and others, making it abnormally easy to dodge and disengage. We can choose to hit "like" or "share" or not; nothing requires us to comment and, at any moment, we can click away from an online interaction.

Even on the phone, we can fake technical difficulties. In person, however, and particularly in places as fragile as peace camps, ignoring someone, especially when they are expressing an opinion aloud in your presence, would be taken as an affront. As the MIT researchers found, it wasn't enough for the less empowered groups to write or speak their minds for their emotions and attitudes to change, they had to feel like there was some kind of meaningful exchange.

We need the same.

Every meaningful relationship in our lives is based on a mutual dependency. Lovers. Parents and children. Neighbors. Yet, our centuries-long aim has been to erase our dependencies. Instead of fading away, though, our dependencies have just shifted: from people to technology. We don't want to be a bother, and we don't want to embarrass ourselves by asking a "stupid question," so we Google. We don't bring our needs to our neighbors, like borrowing a shovel or two cups of flour, we buy our own stuff.

"Love is never abstract," writes Wendell Berry, a farmer and one of America's foremost 20th-century essayists. "It does not adhere to the universe or planet or the nation or the institution or the profession, but to the singular sparrows of the street, the lilies of the field, 'the least of these my brethren' . . . The older love becomes, the more clearly it understands its involvement in partiality, imperfection, suffering, and mortality. Even so, it longs for incarnation. It can live no longer by thinking."

"At its core, technology is a systematic effort to get everything under control."

— Albert Borgmann

As our dependency on others disappears, so do the expected and unexpected intimacies.

Love has arms.

No Sense of Time or Space

My daughter's first movie theatre experience was Dr. Seuss's *The Lorax.* The story centers on a whimsical orange creature who serves as the keeper-of-the-forest. This is the forest of Dr. Seuss's imagination: billowing tufts in cherry red and tangerine. A young man named Once-ler comes to the forest, seeking an entrepreneurial idea that will lead to fortune and fame. He invents the "Thneed," a ridiculously versatile contraption "which everyone needs," and it levels the hills. When he turns around, not a single tree remains, save for one. Years later, a young boy recovers the final seed and, at the climax of the film, the dirt is pulled back and it is planted.

And then, within ten seconds, a tree had bloomed. Seed. Plant. Boom: Tree.

So, when my four-year-old daughter received a terracotta planter and a package of seeds at a birthday party, she expected it to grow. Right now. And she had an expectation, every day, that it would be blooming like that tree she saw in *The Lorax*. When, of course, it shouldn't.

We have forgotten that things take time.

We are annoyed when a package doesn't arrive in time even though we paid for one-day service — forgetting the miracle of the fact that it must pass through 19 hands, 3 trucks, 1 airplane, 6,298 kilometers, and 1 mean dog, to get to us.

This is not to be overlooked.

"In many ways, time and space are the two sides of the coin of human existence," writes Ursula Franklin. "Whatever changes one side will affect the other." So, what happens when we have the ability and the overwhelming desire to manipulate both?

> "I think that as the Internet is such an 'instant' media, there is an assumption that creativity is 'instant' which in some cases can devalue the work of talented, original makers."
>
> — UK artist Karen Ruane. (Read more about creativity and the Internet in Chapter 13.)

Natural Disruption

Perhaps the largest shift in the last century has been the normalizing of disruption. Constant disruption of segmented ideas at hyper-speed — that's where we are today. And, to top it all off, we sound like cavemen. #hashtagsmakeussoundstupid.

Within a moment, we can click from a gut-wrenching image of an emaciated human to a warm, fuzzy image of a newborn baby sitting pretty in a flower pot. Like I did last night. If the imagery is too much, we can click away, feeling we have been informed but incapable of doing anything meaningful with the first image. This phenomenon isn't new to the Internet; it has existed since the dawn of photographs. Nerone and Barnhurst, after studying the news for decades, found that news in fact does not make people smarter, but it does makes the world appear continuously less fair.

And we have never had so much news.

Today, there are more pieces of information on the Internet than grains of sand on the earth, and a million more just popped up on Snapchat, Instagram, Twitter, Facebook, LinkedIn and Pinterest. How many sources can one person keep up with?

Not that many, according to brain health experts. Sherry Turkle, who has studied the psychology of the Net for two decades, says that constant digital interaction impedes our capacity for self-reflection. We can't keep up. Work. Friends. Podcasts. TV shows. What do we have to stem the problem? TIVO.

> Interpretation requires space and time, two things the Internet is designed to nullify.

Knowledge is only acquired after information has been interpreted during communication, says Rembrandt Klopper. And interpretation requires space and time.

Fragmentation, Speed and Mental Health

In *The Real World of Technology,* Ursula Franklin explains how the original modes of communication were always direct, immediate and simultaneous. "The basic process is direct and unmediated, normally one-to-one and of the 'I love you' or 'You still owe me ten bucks' type." Now, one sender can send a message to one person or to a larger group. The mode changes with the addition of a messenger.

"As the sender entrusts the message to an intermediary, the range of communication can be extended, both in time and space . . . the integrity of the message now demands assurance of the truthfulness of the messenger, or communication turns into gossip." In other words, the further away we are from the source, the less can we trust the information.

Franklin continues, "The profound impact of writing, as a technology, lies in the fact that writing allows the physical separation of the message from the messenger or sender." (Even Plato had some choice words about the evil of writing things down.)

"Since the very beginning of writing there have been attempts to assure the authenticity of a given communication," Franklin continues. "Seals or signatures, detailed identifications of the writer or the source of the material . . . As the variety of the messages increased due to the increasing spectrum of senders and receivers, so grew the need to further authenticate, verify and classify the messages."

Read more about trust on- and off-line in Chapter 9: Coming Close.

True communication requires the back and forth of speaking, listening, and then supplying a response informed by what has been heard. This *mirroring* is what therapists insist upon in our most intimate relationships.

You might remember learning in school about the abundance of non-verbal communication exchanged between individuals. It was once calculated that body language accounts for about 90 percent of our interpersonal communication. But it's as if, in a generation, we've collectively agreed that the nonverbal stuff isn't so important. So we text (even to our partners in the room next door).

As I touched on in the last chapter, the visual and tonal cues begin with the face, but they extend to the entire body. Visual communication cues can be found in a person's posture: are they standing, sitting, shrugging, leaning in, leaning back? What are they doing with their hands and arms? Folding them across their chest, placing them in their lap, gesturing wildly, extending a gentle hand in a bid of empathy or kindness?

> "The little machines we now hold in our hands are not neutral. We make them, but they mold us."
> — Professor James K.A. Smith

In most of our interactions this week, we'll never know.

When in Rome

Every era has its epicenter. In the classical era, it was Rome. In the Victorian era, it was London. Today, it is Silicon Valley, USA. And, with the help of the Technorati, we have become obsessed with all things new, no matter how petty. Even Facebook founder Mark Zuckerberg admits most people are more interested in the squirrel in their front yard (or the viral video of a squirrel in someone else's front yard) than the perils of global warming.

In Randi Zuckerberg's book, *Dot Complicated,* she writes about making our online lives as fulfilling as our offline ones. How could that ever happen? Moreover, why are we working so hard to make it so? To ask a still more obvious question, what is the purpose of this technological progress? What higher aim do we think it is serving?

Asking the Right Questions

Every Friday, a teacher asks her students to take out a piece of paper and write down the names of four children with whom they'd like to sit the following week. She also asks the students to nominate one student whom they believe has been an exceptional classroom citizen that week. All ballots are anonymous and the children know that these requests may or may not be honored. Chase, son of the *New York Times* best-selling author Glennon Melton, is in this teacher's class. Melton recounts this weekly practice on her popular blog, Momastery.

After the students go home, the teacher takes out those slips of paper, places them in front of her and studies them: *Who is not getting requested by anyone else? Who never gets noticed enough to be nominated? Who had dozens of friends last week and none this week?*

"You see, Chase's teacher is not looking for a new seating chart or "exceptional citizens,"" writes Melton. "Chase's teacher is looking for lonely children. She's looking for children who are struggling to connect with other children. She's identifying the little ones who are falling through the cracks of the class's social life. She is discovering whose gifts are going unnoticed by their peers. And she's pinning down — right away — who's being bullied and who is doing the bullying . . . It's like taking an x-ray of a classroom to see beneath the surface of things and into the hearts of students."

When asked when she began using this system, the teacher replied: "Ever since Columbine. Every single Friday afternoon since Columbine."

"This brilliant woman watched Columbine knowing that ALL VIOLENCE BEGINS WITH DISCONNECTION," continues Melton. "All outward violence begins as inner loneliness."

And no one needs to be lonely. Not anymore.

4

Dusting Off the Dictionary

Why Definitions Matter

I CAME OF AGE IN A TIME BEFORE GOOGLE.

I remember the day I signed up for my first email address in my senior high school computer lab. Pickings were slim as I combed Yahoo for a pithy name. I settled on "benchfan" (my grade 12 boyfriend chose it; he was in a fledgling band called, you guessed it, *Bench*) thinking nothing would become of this strange computer mail. As it turned out, "benchfan" would serve me for nearly a decade, through several years of university and beyond.

I Love Email

My first email experiences were astoundingly memorable. An early romance budded by way of after-date messages. I recall the thrill of sitting in front of our downstairs family computer, heart pounding, waiting for that assured good night message. Back then, getting an email was like Christmas morning.

How things have changed. Today, email is our great nemesis, the volume crushing. With 200+ messages a day, "Inbox Zero" — the mythological pursuit of processing every message every day — is as fleeting as the tinker fairies of my daughter's imagination: cute, but unreal.

Our earliest computer technologies, once boxy stationary items, have shrunk to the size of a palm and can travel with us everywhere. Where, at one time, people demanded raises when an employer required round-the-clock access via a BlackBerry, now this kind of availability is an unwritten rule (and employees have to provide their own device!). Employers are to blame for their lack of boundaries, and we are for our complicity.

We are always on, never off. Constant access isn't a blessing anymore.

Like most 21st-century humans, I am burning the candle at both ends. When I feel the world spinning too quickly and have a moment's respite, I don't look up, I look online.

My children are little. Every night I am awakened multiple times to feed a baby, rescue a sippy cup, or comfort a little one after an unsettling dream. When not writing, I spend my days bent low cleaning linoleum, shoveling sand and mopping messes. It's beautiful and thankless work. When the last child is finally quiet in their bed, it is all I can do to fling myself on the comforter and type in the words: "Netflix." But, as grace would have it, I am not a single mother. I have a partner, and our relationship requires nurturing, especially in these exhausting years with young charges. Of course, there are nights when we extend each other permission to check out and watch a show. But our relationship feels the gap, and if we are not careful, that gap widens into a canyon. Our consumption requires intention.

How do you view the Internet? As a tool? A looking glass? An escape route? We use the Web for many reasons but, increasingly, especially with the adoption of smartphones, our engagement looks more like compulsion than intention, and compulsion, by definition, "is an irresistible urge to behave in a certain way *against* one's conscious wishes."

"Technology is not just a tool, it's an inducement," says Albert Borgmann. "And it's so strong that for the most part people find themselves unable to refuse it. To proclaim it to be a neutral tool flies in the face of how people behave."

I tap the Facebook app because, well, that's what I do. This indifference to an obviously compulsive behavior lets a day slip quickly by.

Defining Our Approach

How does the Internet serve you? Is it connecting you — truly — in ways that bless and enliven your life and the lives of others? Is it a tool that helps? Do you learn and act more because of it? Is it displacing burdens that you should not want to be rid of? Are your engagements online helping make you who you want to be? What can we learn from mining our past about relationship, human nature, efficiency and the imagination? Is more information making us dumber? People once had less to read, fewer things to play with (stick and box), which seems to have led to a wider imagination. Teens wrote epic novels (*Frankenstein,* anyone?) and engaged in deeper conversation. Is more and faster information making us smarter or is it working in the reverse?

In the 1980 film *The Gods Must Be Crazy,* we follow a small and happy tribe as they hunt, cook, gather, and eat. All the while, they play and laugh, chattering in a wide range of melodic "kloks." In contrast, a few miles away, a labyrinth of highways stretch as far as the eye can see, smog hangs thick in the air as we watch a woman, dressed in a robe and curlers, jump in her car to drive 20 feet to deliver some letters to a mailbox before reversing at record speed back to her driveway.

Which seems more civilized?

Our technologies, while aiming to make our lives easier, have, in effect, made them more complicated. And we carry our complications with us. But, as the brief history in Chapter 2 showed, this has not been an overnight story. It has been a long progression, one with many creators, advertisers and willing adopters. And it is a global story.

How we view our computer technologies directly impacts how we value and interact with them. If we desire to cultivate a more meaningful existence in our media-saturated world, we must define our approach. Since the late 1990s we have embraced email, cell phones and Google with wild abandon but our habits and disciplines have not caught up. Clarifying our view of the Internet will help us to know when to stop.

> "Civilized man refused to adapt himself to his environment. Instead, he adapted his environment to suit him. So he built cities, roads, vehicles and machinery, and he put up power lines to run his labor-saving devices. But somehow he didn't know when to stop. The more he improved his surroundings to make his life easier, the more complicated he made it."
>
> — *The Gods Must Be Crazy*

> "Joe has a small PC fix-it business near my town. Joe is from Ghana. I was asking him about life there. He said, 'People are connected to each other there. Here, people are connected to machines.'"
> — Leslie, Massachusetts

> The fundamental problem is volume.

A Volume Vortex

I have already mentioned the little known fact that there are more pieces of online information than there are grains of sand on Earth. Sometime in 2012, God passed the baton to Google.

The Internet is many things. At its best, it is an unprecedented source of information and a communication enabler. At its worst, a narcissistic mind-number and a portal for predators. Above all, the Net is a *volume vortex.*

"There's just too much of everything out there," laments actor Suzanne Pringle from Montreal. "I feel as though I'm living in a landfill teeming with objects and ideas, and millions of people are scrambling to make piles of things to make their mark, and quickly, because in a moment, more ideas and objects will be dumped by the tons on our heads."

Joshua Fields Millburn echoes Pringle's sentiments. Half of the blogging duo The Minimalists, he recently disconnected his high-speed Internet at home. Millburn, an essayist by profession, realized his productivity was waning. He felt he could do more meaningful things than spend spare time online — he swapped his go-to online fillers for exercise, writing and strengthening his existing relationships.

"This doesn't mean I think the Internet is evil or bad or wrong (obviously it's not) . . . The Internet is not evil, just like candy is not evil. But if your entire diet consists of candy, you get sick and fat fairly quickly. Thus, I don't keep bags of candy at home, just like I don't keep the Internet at home anymore either." Instead, he posts, reads email and even schedules frivolous Googling while connected to Wi-Fi for an hour at his local coffee shop.

Having social media on your handheld is like living on a cotton candy mountain. With every juicy morsel of information available at our fingertips, abstaining is a near impossibility, even for the most disciplined among us. (Even though I am completely immersed in this research, it took me about three weeks and a Herculean load of self-talk, to get me to delete my Facebook app, *again.*)

The information on the Internet, like a vortex, is a mass of whirling air: ever-changing, ephemeral, entirely ungraspable. In the time it took to read the last sentence, Facebook added 118,000 new updates and the Huffington Post added 18 new stories. We can't keep up.

We haven't figured out what to do with it all yet.

But Nicholas Carr, author of *The Shallows: What the Internet Is Doing to Our Brains* tells us that our focus on the medium's content — the words, the apps, the websites — blind us to its deeper effects.

"We're too busy being dazzled or disturbed by the programming to notice what's going on inside our heads. It's how we use it that matters, we tell ourselves. The implication, comforting in its hubris, is that we're in control . . . What both enthusiast and skeptic miss is . . . that in the long run a medium's content matters less than the medium itself influencing how we think and act . . . Media work their magic, or their mischief, on the nervous system itself."

How often have you flipped open your laptop to look something up only to find yourself, 20 minutes later, on some random site — unable to remember why you opened your browser in the first place? That's the vortex. So, while disconnecting our Internet at home may be step in the right direction, we must see the forest for the trees.

Our computer technologies have rewired us. Volume management is only a step in the right direction. We are not the same people anymore.

Every new medium changes us.

Experts say that multitasking is a fallacy. One cannot do more than one thing at a time; instead, our mind rapidly shifts from one thought or task to another. As I type these words, I have six books open on my desk and 19 tabs open on my MacBook (really, I just counted). My mind, as you might expect, is switching gears many times a minute. I've now closed the lid on my laptop and picked up one of those books: *Smarter Than You Think* by Clive Thompson. As I settle in, somewhere along the fourth page, something shifts. I begin to feel the wheels moving in my head, drawing my thoughts forward, like a train's gentle rocking, picking up pace.

Jane Austen's *Mansfield Park* is a complex, challenging novel read by millions. A recent collaboration between Stanford neurobiologists and English postdoctorate student Natalie Phillips suggests that complex novels such as *Mansfield Park* can activate key brain areas. The group of researchers at the Stanford Center for Cognitive and Neurobiological Imaging took brain scans of several literary Ph. D. students as they read a chapter from the book. First they were asked to read for fun, then more critically. Critical reading created a significant shift in brain activity patterns on MRI scans, increasing the activity of the prefrontal cortex, which is responsible for executive function. Executive function is responsible for more than just attentive reading; this brain function helps moderate how you divide your attention, use your working memory, and generally direct your brain power. It plays a powerful role in decision-making.

Avid long-form readers, especially those who began these types of intellectually challenging habits early on, have more cognitive reserve. Now, consider this: the average person who conducts 90 percent of their reading online spends no more than two minutes on a webpage.

> "In the 'cult of the new' (which is the genius of capitalism) novelty is fleeting as pixels on a screen. Perhaps a better guide for the future is an elder medium such as print. It is rugged and soft at the same time. It's patient and demanding."
>
> — *Geez* magazine

As I have engaged more and more online, I have shared Nicholas Carr's uncomfortable sense that something has been tinkering with my brain, remapping the neural circuitry and reprogramming my memory.

I can remember the first time I wanted to Control Z something in real life. I was painting a canvas when, all of a sudden, I made an unfortunate brushstroke. "Undo!" my mind raced, "Control Z! Back up, back up!" But, instead, I had to embrace the mark and carry on.

Behavioral psychologist Ellen Langer got the idea to study mindfulness and mistakes in a similar moment. She says, in interview with the *Harvard Business Review*, "[I was painting and] I looked up and saw that I was using ocher when I meant to use magenta, so I started trying to fix it. But then I realized I'd made the decision to use magenta only seconds before. People do this all the time. You start with uncertainty, you make a decision, and if you make a mistake, it's a calamity. But the path you were following was just a decision. You can change it any time, and maybe an alternative will turn out better. When you're mindful, mistakes become best friends."

It's not surprising to think we should be able to Control Z our lives when we spend the lion's share of our lives in a space where such things are possible. Not even Marshall McLuhan, the father of communication theory, could have foreseen "the feast that the Internet has laid before us: one course after another, each juicier than the last, with hardly a moment to catch our breath in between bites," as Carr vividly describes.

Ninety percent of adults and teenagers own smartphones, and the number is rising steeply every day. Thanks to our digital technologies, our attention spans have decreased from a span of 12 minutes to 5 in just a decade. If a website doesn't grab us in 30 seconds or less, we click away.

The new media has restructured our experience of the old media, says Read Schuchardt. It's under the conditions of multitasking, where you get to be two people, two places at the same time, that gives us the thrill that all media and all portable media allow us to do whatever we want, whenever we want. "The problem is, when you have been given the pleasure of multitasking, even the divine pleasure of reading — no matter how great the content — becomes mono-tasking, a kind of punishment."

We're remapping our minds, byte by byte.

Everything Is Amazing, and No One Is Happy

In a diatribe delivered on *Late Night with Conan O'Brien,* comedian Louis C.K. explains why he dislikes the culture of smartphones and why he would never get one for his kids.

He begins by suggesting that smartphone usage is the reason kids today are meaner: "I think these things are toxic, especially for kids . . . they don't look at people when they talk to them and they don't build empathy. You know, kids are mean, and it's 'cause they're trying it out. They look at a kid and they go, 'you're fat,' and then they see the kid's face scrunch up and they go, 'oh, that doesn't feel good to make a person do that.' But they've got to start with doing the mean thing. But

> "You need to build an ability to just be yourself and not be doing something. That's what the phones are taking away, is the ability to just sit there."
>
> — Louis C.K.

when they write 'you're fat,' then they just go, 'mmm, that was fun, I like that.'"

From there, C.K. moved on to expound on the larger issue: The negative emotional effect that smartphones have on grown-ups. While C.K. agrees that smartphones can help create a sense of community, he believes that therein lies the problem: "You need to build an ability to just be yourself and not be doing something. That's what the phones are taking away, is the ability to just sit there. That's being a person . . . And sometimes when things clear away, you're not watching anything, you're in your car, and you start going, 'oh no, here it comes.' That *I'm alone* — it starts to visit on you." So, that's why we text and drive, he says.

"I look around, pretty much 100 percent of the people driving are texting. And they're killing, everybody's murdering each other with their cars. But people are willing to risk taking a life and ruining their own because they don't want to be alone for a second because it's so hard."

Finally, C.K. brings it all together with an anecdote about the time he was in his car listening to Bruce Springsteen's song "Jungleland." He was alone and as he listened to the words he could feel sadness coming over him, a deep, deep sad, and his first impulse is to grab his phone and write "hi" to 50 people to keep the feeling at bay. But, instead, he decides to be in it. To let the emotions wash over him, to give himself up to the sadness.

"And I let it come, and I just started to feel 'oh my God,' and I pulled over and I just cried like a bitch. I cried so much. And it was beautiful. Sadness is poetic. You're lucky to live sad moments.

"And then I had happy feelings. Because when you let yourself feel sad, your body has antibodies, it has happiness that comes rushing in to meet the sadness. So I was grateful to feel sad, and then I met it with true, profound happiness. It was such a trip.

"The thing is, because we don't want that first bit of sad, we push it away . . . [so] you never feel completely sad or completely happy, you just feel kinda satisfied with your product, and then you die. So that's why I don't want to get a phone for my kids."

> If we want to tread new and deeper paths, we must chart a different course.

Our computer technologies give us the ability to send a message, watch a show, play a game, buy a product, read something new. The problem with high-volume media is that we are bombarded, fragmented, addicted: running from dopamine-hit to dopamine-hit and, as a result, our emotional regulation is skewed. We are fragmented people.

And fragmented people make the best consumers.

Haste Makes Waste, or Fragmentation Leads to Consumption

We have a term for it: "Retail Therapy." Where once we had to get out of our pajamas, lug ourselves out the door, and hit the mall, now, when we feel sad, we hit eBay. *Voilà.*

But how did this consumer culture sneak into our way of thinking?

In his book, *Being Consumed,* William Cavanaugh argues that we, regardless of a mooring religion or not, have been greatly shaped by and captivated by our consumer culture — a culture which says joy and happiness and the good life can be found when we buy certain products. And this way of thinking seeps into our lives in a thousand different ways, through the advertisements we consciously and unconsciously absorb on- and off-line.

"In consumer culture, dissatisfaction and satisfaction cease to be opposites, for pleasure is not so much in the possession of things as in their pursuit," he writes. "There is pleasure in the pursuit of novelty, and the pleasure resides not so much in having as in wanting. Once we have obtained an item, it brings desire to a temporary halt, and the item loses some of its appeal. Possession kills desire; familiarity breeds contempt . . . The consumerist spirit is a restless spirit . . . desire must be constantly kept on the move."

And we've never been moving faster.

> "When we rush, we are more likely to consume because we are ignoring the little voice asking us if we really need this new thing. Impulse buying is what corporations depend on."
>
> — Michael Schut

A Money Maker

The creators, investors and techies in Silicon Valley are motivated by two things: the thrill of invention and money. But, make no mistake, the latter is king. The "cult of the new" reigns supreme. Our technologies are dead in the water after only a couple of years. Upgrade we must, or get left behind. And our Bay Area friends are laughing all the way to the bank.

"We (as app makers) want them to be addicting. Like a potato chip manufacturer, we try to put just the right crunch and the perfect amount of salt so you can't help but have just one more. We want you to get addicted. It puts the potato chips on our table," says mobile app developer Jeremy Vandehey.

Money is made simply: by keeping us clicking.

The world is full of unpredictability and change. In order to keep us interested, the Internet must mimic this unpredictability. It's why every social media site has several columns with real-time updates, why every news site full of advertising posts new stories every few minutes. (Save for Wikipedia and a few other ad-free websites, the majority follow this format.)

The tweaks keep us interested, stopping us from moving on to other things. Where a kid playing naturally moves on to something new, we, instead, get stuck.

The tech makers and earners such as David Sarnoff, the media mogul who pioneered television at NBC in the 1950s, are quick to move the blame for the ill effects of a new medium away from the technologies and onto the listeners and viewers. Nicholas Carr recounts Sarnoff's speech at the University of Notre Dame in 1955, perhaps the glory year of TV: "We are too prone to make technological instruments the scapegoats for the sins of those who wield them. The products of modern science are not in themselves good or bad; it is the way they are used that determines their value."

> It's repeatable behavior with sporadic positive reinforcement. We check email 18 times a day in the hope that one of those times we'll find an exciting message. What could we have been doing those other 17 times?

So said the creators of nuclear technology. As lovers, citizens, neighbors and parents, we know *motive matters.*

"It is very difficult to step out of the immediacy of the Internet because we believe we have control over these technologies, but the truth is we make our technologies, but they remake us in their image and for their purposes," says media expert Read Schuchardt.

And, by design, their purpose is the bottom line.

Facebook, Gmail and thousands of apps are free. But are they really? With every email, post and click, preferential ads are doctored and projected onto your screen and their message is universal: *You are not enough.*

This is no new trick; it's been the unspoken line of advertisers since the dawn of classifieds — or older still. Remember Adam and Eve in the Garden of Eden? The snake fed them the exact same lie.

You are not enough, eat this fruit.

You are not enough, buy this stuff.

The Internet is monetized at every turn. To advance in the ultra-popular online game Candy Crush, you can choose to send ads to your Facebook friends, unwittingly promoting the game dozens of times a week (as my cousin does relentlessly, much to my dismay). Advertising is in adapt-or-die mode as the Internet changes by the nanosecond. But if you think it's letting up, you're wrong.

Consider this:

> "You're getting drowsy behind the wheel. Your phone detects your drooping eyelids and sounds an alarm. Jolted awake, you pull over and see your phone is guiding you to the nearest place to get a wake-me-up fix — the nearest, that is, in the chain of coffee shops that built the watchful app.
>
> You've just finished a run, and as you log your workout data into a fitness app, up pops a reward: a coupon for a free sports drink.
>
> Your baby is irritable. Her high-tech diaper sends an alert to your phone that explains why, courtesy of your favourite diaper-cream brand.
>
> Welcome to the brave new world of advertising."
>
> — Susan Krashinsky, *Globe and Mail*

The final frontier of computer technologies is to seamlessly integrate with humans (before replacing them entirely). If we continue on the same trajectory, wearable computers (Google Glass and the like), "smart" clothing and diapers, and computerized homes will become commonplace in a matter of years.

Behavioral economics tell us that there is a psychological cost in choosing what we buy. Our technologies give us practically limitless choice, but does choice make us happier? In what is now conventional wisdom, Sheena Iyengar, in her breakthrough TED talk "The Art of Choosing," tells us "no." Just as children, in ritual and routine, need the absence of choice to emotionally regulate, so do we.

Humans thinking of themselves as machines has a long history, dating back to the classical era, explains Katherine Hayles, author of *How We Became Posthuman,* in an interview. "Since World War II and the development of intelligent machines, this tendency has greatly increased . . . Think of all the everyday expressions that now equate human thought with computers: 'That doesn't compute for me'; 'my memory is overloaded'; and my favorite, drawn from [Sherry] Turkle's account of the world of hackers: 'Reality is not my best window.'"

If we were to take a careful examination of our trajectory, where would we be headed? What is the purpose of this technological progress? What higher aim do we think it is serving?

Real Life Is Not My Best Window

The Internet began with a vision of open source information. The "democratization of the Internet" was an early catchphrase for hackers and early Web visionaries. Sherry Turkle was one of those early dreamers.

In her 1996 TEDTalk she optimistically pronounced: "Those who make the most of their lives on the screen come to it in a spirit of self-reflection." She believes the same is true now, but something has shifted along the way.

She has recently learned that people have been mastering the skill of making eye contact while texting. Apparently it's difficult, but it can be done. Why does this matter? Turkle believes it matters because we're setting ourselves up for trouble.

In her book *The Second Self,* Turkle writes about how computers are not tools as much as they are a part of our social and psychological lives. "Technology," she writes, "catalyzes changes not only in what we do but in how we think."

For more than 20 years Turkle has been studying the psychological and societal impact of such "relational artifacts" as social robots, and how these and other technologies are changing attitudes about human life and living things generally. More and more we are devaluing authentic experience in a relationship.

When we feel like no one is there to listen to us, we want to spend time with our technology to fill the gap. Turkle has found that, increasingly, people wish for an advanced version of Siri (Apple's voice-command application) that will be a friend who will listen when others won't.

Our emerging technologies are changing what we believe about being human.

Turkle conducted research in one nursing home where a woman, who had just lost a child, was interacting with an "empathetic" robot that seemed to respond to her. The staff was impressed. But Turkle called it "one of the most wrenching, complicated moments in her 15 years of work." "Have we so lost confidence that we will be there for each other?" she laments.

Turkle says we are creating technology that give us the illusion of friendship without the deeper demands and complexities of relationship. In her view, we are being offered three fantasies:

1. That we'll have attention everywhere.
2. That we'll always be heard.
3. That we'll never have to be alone.

"This relationship, this constant connection, is changing how people think of themselves, it's shaping a new way of being . . . I share, therefore I am." Turkle believes we need to cultivate a capacity for solitude. "If we're not able to be alone, we're going to be more lonely. And if we don't teach our children to be alone, they're only going to know how to be lonely."

An Escape Route

The greatest meaning in our lives is found in love. In being real. In fact, the ancient Greeks called God *the really real.* And the Internet can help us to love; it can give us opportunities to be candid, to confess; it can help us in our work, bridge relationships, give us opportunities to help others.

But it can also become a crutch, a compulsion, an escape route. With our ever-increasing use of online technology, the idea of community is shifting profoundly — whether it be family, friends or relationship within a local faith community. While the Internet has enabled the recluse, the elderly, the sick and the disabled to connect to others online, it has created an unnecessary hurdle for the rest of us.

Far-away relationships require mediation: email, phone, letters — they all bridge distances beautifully, in their own way. But, when it comes to closer to home relationships, we have mediated what does not need mediation. I don't need to text my neighbors; they live ten steps away. I don't need to phone my husband; he's downstairs. Our ubiquitous online media have made the ease of mediation irresistible.

Ease in itself is not a problem. We all need downtime, respite, relief. Ease at all costs, thumbs trained on our escape route, is.

The middle man — our computers, tablets and smartphones — allows us, for the most part, to remain at arm's length. And, in the process, we, the unlonely, have grown lonely. But there is a simple way to counter it. Whenever you have an opportunity to see people *in person* — a dinner, coffee date, street party — GO. Feel yourself withdrawing, saying "no" to real-life events in exchange for another night with Netflix? Remember: real life is your best window.

The Ultimate Tool Box: Using the Internet to Fill a Real Need

Acclaimed behavioral-science writer Winifred Gallagher explains that the best advice on how to live in a world of potentially limitless distractions boils down to two words: *selectivity* and *moderation.*

Considering the brain's limited energy, Oshin Vartanian, a psychologist at Defence Research and Development Canada, advises asking some questions before engaging with something new:

- Why should I take this up, if my daily scripts are doing a good job for me?
- Why exactly do I need another gadget?
- It will incur certain mental costs, so where will those resources come from?

Ask the right questions: Is it a tool that helps? Minimalist Joshua Fields Millburn plans his usage of the Internet, harnessing it for its good. In order to cultivate a richer inner and outer existence, we must approach our lives with the same intentionality. Viewing the Internet as a tool helps lead us away from compulsion to more intentional consumption and greater balance in our lives.

Technology is not neutral, but the answer isn't to stop using computers and the Internet entirely. That would, in some ways, be easier. Indeed, that is what I found. Thirty-one days entirely offline was far simpler than navigating my day-to-day practice since. The answer is difficult: we need to figure out what's most important and inspiring for us, and then subordinate our use of technology accordingly.

"When I ask my computer to remember things for me, my memory is neglected," writes Aiden Enns, publisher of *Geez* magazine in Winnipeg. "When I ask my computer to look up the definition of a word for me, my dexterity diminishes. Moreover, my predilection for ease is nourished — I begin to expect and rely upon that which is easy. When I devote an evening to watching my computer with my lover by my side, regardless of the movie we've rented, I have left the room and entered a better place that is, unfortunately, less real."

He continues: "That is what is bad about technology as we commonly think of it: even though we are more productive, connected and entertained, at the same time, we ourselves become less functional as sentient creatures. As beings capable of humour, poetry, whimsy, devotion, compassion, grieving and surrender, when we inundate ourselves with technology, we lose our focus and begin to act like machines."

Like Enns, I'm not rejecting technology; I'm aiming for a mindful approach: if it's good for my character, soul and community, I'll judiciously engage.

Chapter Challenge:

Think about the things you need to do online today. Now write them down:

____

____

____

____

____

When you choose to go online, complete the list and get off. Tomorrow: rinse, repeat.

5

Introduction to Part Two

Presentness

i thank you God for this amazing
day: for the leaping greenly spirits of trees
and a blue true dream of sky; and for everything
which is natural which is infinite which is yes

— e. e. cummings

Chapter Challenge:

Read this poem aloud. Note in the margins how it felt.

"I LOVE EVERYTHING THAT'S OLD: old friends, old times, old manners, old books, old wine." You'll find these words from Oliver Goldsmith chalked on a coffee-of-the-day board steps away from the Regional Assembly of Text — a small paper emporium that would make Ned Ludd proud. Here co-owners Rebecca Dolen and Brandy Fedoruk, grads of Emily Carr University of Art + Design, stand behind the counter of their store, a wall of cast-off industrial filing cabinets behind them; they are assembling cards and packages with meticulous care. Their space is notably lacking any computer technology — or even a phone. Orders are written up by hand on rubber-stamped receipts. It's a stark contrast to Vancouver's noon-day bustle streaming by outside, moments from the corporate homes of Electronic Arts and Lululemon, and mindfully so.

Quiet spaces like these are becoming increasingly popular, a refuge from our perpetual state of information overload.

There's no question that technology has overrun our lives. Over the past century, the world has welcomed technological "progress" with arms wide open, and we're living with the clicking, dinging, anxiety-inducing deluge of it. But a creative backlash is underway, helping human beings cope with the avalanche of data that passes in front of most of us every day through the use of computers and smartphones.

Slow food, the back-to-the-land movement and groups like letter writing clubs are being formed by a new subculture: the 21st-century Luddite. They wield fountain pens and notebooks; some spend a mere hour per week checking their email — at the public library.

Dolen and Fedoruk think this movement is more than a blip on the technological continuum.

"We started the letter writing club right off the bat because we wanted to have an ongoing community event. There have been a few hardcore regulars, but 80% are new people each month. We started with five to ten people and now regularly have 20 to 30."

There's a universal sense that something must be done to rope the nodes in. But what? We can't all pack our bags and head for the hills, or can we?

As a student at MIT, Eric Brende became a critic of modern technology. He left the bookish hollow of Boston and immersed himself in an Amish-like community in middle America. He chronicles the appeal of a slower life in his book *Better Off: Flipping the Switch on Technology.*

"Time moved more slowly . . . we had more *of* it . . . in the absence of fast-paced gizmos, ringing phones, alarm clocks, television, radios, and cars, we could simply take our time. In being slower, time is more capacious. The event is only in the moment. By speeding through life with technology, you reduce what any given moment can hold. By slowing down, you expand it."

In the span of a generation, items like the typewriter have pretty much disappeared from the communications lexicon; but, thanks to a growing group of modern-day Luddites, they are making their way back.

And so, on Canada's west coast, the Assembly of Text tries to create some space with their simplistic, low-tech aesthetic. Every month, the popular letter-writing club is facilitated by a handsome collection of 30 some-odd typewriters. Filmmaker Andrew Blicq was drawn to this scene for a Merit Motion Pictures film exploring the impacts of information overload.

"There really aren't many groups who have gone completely off-line," Blicq says of his choice to film the gathering for *Our CrackBerry World,* a CBC Television documentary.

The concept of "information overload" can be traced back to Diderot, though he didn't quite use that term. In his *Encyclopédie* (1755) he wrote, "As long as the centuries continue to unfold, the number of books will grow continually, and one can predict that a time will come when it will be almost as difficult to learn anything from books as from the direct study of the whole universe. It will be almost as convenient to search for some bit of truth concealed in nature as it will be to find it hidden away in an immense multitude of bound volumes."

Diderot was right. Today, myriad reasons are drawing people to reduce their amount of input, whether it be blog feeds, audio books, television or email.

For one letter writer, the appeal of a slower form is its mindfulness: "You have to think about what you're writing because you can't erase it."

Older technologies, such as the typewriter and the fountain pen, require forethought. There's also a sense of humanness, of the real and the unmediated.

"The handmade is coming back because everything is too standard now," reflects Rebecca Dolen. "With the typewriter, it's not nostalgia. People are too young for it to be nostalgia. We always had one around the house but didn't use it. There's just more personality with the typewriter. With mistakes and everything, it feels like it's really you."

"You can edit too much on email. Maybe there's a release with the typewriter and the handwritten forms," chimes in Fedoruk, both nodding their agreement.

"People are back into the letterpress big time," echoes Fedoruk. "W2 [the global media arts center in Vancouver's downtown Eastside] started a letterpress studio from the get-go. Everybody wants a typewriter. We get 100 requests a week." Inquiries are sent on to Art Polsons, a Terminal City fixture, who refurbishes and re-sells scores of Olympias and Remingtons to a growing clientele.

It's the messy mindfulness of the handwritten note and the therapeutic clack of the typewriter that are drawing a new breed of writer. To many, it's a form of escapism, drawing one's mind away from the distraction and interruptions of our 140-characters-or-less lives.

"There's this sense that, especially with text messaging and BlackBerry chat, that it can't wait. Messages have to be answered immediately," comments Blicq in an interview from his summer cottage, aware of the irony as he finishes up film production from his remote hideaway.

"No one's going back to the Smith-Corona [typewriter]," he goes on, "but we're also sick of using the BlackBerry in the bathtub and the car. Our work weeks have stretched from 40 to 70 hours with the introduction of the smartphone. There's no question something's missing from our lives.

"It was ironic to be asked [by the CBC] to direct this film. Everyone in the TV/film industry has a smartphone, and Internet and computers are everywhere. We all have a problem with juggling. Once we hit the streets of Toronto and Vancouver and asked people about technology, the universal answer was: "Yes, it's too much." But we all like it, we'd be lost without it. Also, I think that to be a part of the global dialogue you have to plugged in to some degree.

"Technology is not going away, the genie is out of the bottle. But the big question we need to be asking ourselves is: 'Is it going to manage us or are we going to manage it?'"

A few months ago, a little boy stumbled into the Assembly of Text with his mother. He sat down on the couch at the front of the shop and began to tap on the resident typewriter. Stunned at the words forming on the page in front of him, he announced with glee:

> "Mom, the letters go right onto the paper!"

Maybe what we all need is a little bit more of that.

6

Why Fast from the Internet?

Finding What Sustains

Soon silence will have passed into legend. Man has turned his back on silence. Day after day he invents machines and devices that increase noise and distract humanity from the essence of life, contemplation, meditation . . . tooting, howling, screeching, booming, crashing, whistling, grinding, and trilling bolster his ego. His anxiety subsides. His inhuman void spreads monstrously like a gray vegetation.

— Jean Arp

God rest us. Rest that part of us which is tired. Awaken that part of us which is asleep. God awaken us and awake within us. Amen.

— Michael Leunig,
When I Talk to You: A Cartoonist Talks to God

IMAGINE YOU ARE SITTING WITH YOUR FRIEND at the Earwax Cafe in Chicago's Wicker Park neighborhood. You're mid-conversation when he realizes he's forgotten to send an email, so he grabs his phone. You've come empty-handed, so you sip your latte and people watch, being careful to avoid eye contact with the passersby.

We've all been there: the halted conversation because someone checked out to use their smartphone. It's not a great feeling.

It's moments like these that capture the profound global shift we've experienced in the few years since the first smartphone was brought to market. Deep down, we believe we have control over our mobile technologies; the truth is we make our technologies, but they remake us: the way we see the world, the way we spend our time and the way we value and relate to others.

And, according to Google, this is just the tip of the iceberg.

We Live in the World of Internet Everywhere

The future, according to Ben Gomes, Google's Vice President of Search, is for the search monolith to be "present everywhere." Consider this proposition: The very idea of presentness speaks of specificity — a particular time and place. But Google aims to offer its worldwide users omnipresence. Already, the company is universally recognized as the world leader in searching for information. It handles around 90% of Internet searches. When we want to know something, most of us turn to Google. But, as Ian Burrell reports in the UK's *The Independent,* "It wants more — it wants to become our constant companion."

In 2013, Burrell was invited to try Google Glass, a featherweight technology that fits like a pair of glasses that enables you to walk down busy streets receiving helpful facts — without needing to take your mobile phone from your pocket. By uttering the magic words, "Ok, Glass," you are offered a range of options: ask a question, take a picture, record a video, get directions, send a message, make a call, or make a video call.

"The potential is enormous," says Burrell, "a proud mum could film her son taking a penalty kick in the football match while dad, abroad but connected through Google Hangout service, could watch the action live through his wife's Google Glass."

While the father may relish this opportunity, the son, on the other hand, may not experience this as presentness, instead longing for his father's physical presence.

If we have all information and connectivity available to us at all times, will we have the ability to ever be truly present. Do we now?

In the episode "The Entire History of You" on BBC's TV series *Black*

Mirror, the writers take us to an alternative reality where most people have a "grain" implanted behind their ear which records everything they do, see or hear. This allows memories to be played back either in front of the person's eyes or on a screen, a process known as a "re-do."

Would a community rejecting something like the "grain" create a new kind of Amish? The Amish abstain from new technologies in order to honor the God they serve, out of reverence for the holy omnipresent spirit they have chastened their lives to, and to keep their commitments to community. In the Garden of Eden, Satan tempts the first man and woman with all knowledge. Would the great spiritual act of the 21st century be to refuse this knowledge? Is this knowledge too much for us?

Problems in the Most Wired City on Earth

Parents in Seoul are discovering that being the "most wired" city on the planet has its consequences. According to a *New York Daily News* article, "South Korea battles smartphone addiction," more than 80 percent of South Koreans aged 12 to 19 owned smartphones in 2012 — double the 2011 figure. Nearly 40 percent of them spent more than three hours a day tweeting, chatting, or playing games — despite attempts by teachers to confiscate all devices at the beginning of the day and return them when classes were over. An annual government survey estimated that nearly 20 percent of teenagers are "addicted" to smartphones.

Addiction was defined by a number of criteria, including anxiety and depression when separated from a smartphone, a repeated failure to cut back on usage time, and feeling happier using smartphones than being with family or friends.

"Many young mothers nowadays have their babies play with smartphones for hours to have some peace at home, which I think is really dangerous," states Lee Jung-Hun, a psychiatrist at the Catholic University of Daegu. "The younger you are, the easier it is to become dependent."

South Korea is painfully aware of the tragic results that can flow from Internet dependence. Police in 2010 arrested a couple who let their three-month-old starve to death while they obsessively played an online game — about raising a virtual baby.

This is a stark example, but it reveals how the Web, when not used mindfully, can become an entrapment.

The Examen of Consciousness

With some exceptions, our websites and apps do not encourage reflection; instead they urge us to consume: more email, more goods, and more information. Screen time, when used at length mindlessly, falls firmly in the *life-taking* column. But, the truth is, consumption is not our great and final goal. We were made for more.

The Examen of Consciousness can help lead us there.

The Examen, developed by Saint Ignatius Loyola (1491–1556), founder of the Jesuits, is a daily practice of spiritual reflection intended to help a person become aware of the presence of God in their life. At the end of each day, a person asks themselves: "For what today am I most grateful?" and "For what am I least grateful?" Put another way: *what today was most life-giving and what was most life-taking*?

St. Ignatius believed that the key to a healthy spirituality was to find God in all things and work constantly to gain freedom in our lives in order to cooperate with God's will — that meant seeking the good and the life-giving, in the belief that God leads us to do more of those things that bring delight and peace, fewer of those that bring heaviness and less of that which *takes* life. Over time, this centuries-old practice of reflection can be a powerful tool.

Chapter Challenge:

Take a moment now to answer: *What today was most life-giving and what was most life-taking?* (Your answers will help you with the challenge at the end of this chapter.)

____

____

____

> The Examen encourages us to ask two questions: *What today was most life-giving* and *what was most life-taking?* These reflections can help us cultivate an existence that centers on life.

The Whole World Fasts

Every religion in the world has a form of fasting — seasons set apart to better see and hear the Spirit whom they follow. Ramadan. Lent. Navratri. Yom Kippur. These are times for abstaining from the things that disconnect us from the divine, from the true, from the things which give us life. In every culture, distractions and vices are set aside for a time to reach a higher good such as health, clarity in prayer, or intimacy with others.

In the practice of *mannat* Sikhs give something up in order to ask for something.

"Should I ever stop fasting from all that numbs, dulls and deadens me to life?"

In her *New York Times* reflection, Catholic María de Lourdes Ruiz Scaperlanda reflects on the season of Lent, a 40-day fast leading up to the remembrance of the death and resurrection of Jesus in Christian traditions: "The question for me is not whether there's a point to giving things up during Lent, but whether I should ever stop fasting from all that numbs, dulls and deadens me to life, all of life, as it is today — the good and the bad. Fasting makes me willing to try."

Better Off

A number of months into life on his adopted low-tech homestead, Eric Brende dug into the pages of *The Education of Henry Adams,* a 700-page tome that had intimidated him for years.

"Tonight, to the flicker of a kerosene lamp, I made inexplicable, rapid progress," he recounts in his book, *Better Off.* "In the modern university, with its rapid turnover of assignments and fast-paced technology, the human brain is treated as just another processing device and is expected to keep pace with electronic blips. But Adams's thought, ponderous and discursive as it was, could not be summarily ingested. He had lived with a culture whose movements were still largely limited by the speed of horses; the ambling cadences of his writing preserved this pace . . . This was the secret: to grasp his meaning, you had to be living it. Not merely your thoughts, but your various daily duties . . . had to fold together in a quiet rhythm, an interconnected unity."

It was my friend Matthew who recommended I read Brende's book.

Matthew is the type of friend that knows the scientific names of trees and birds and isn't ashamed to tell you. He always asks if I'm reading anything interesting. He's the kind of friend who will pull you into a kayak rather than meet you at a coffee shop, and the kind who'll read *The Hobbit* aloud to your toddling one-year-old regardless of her level of interest. He's also obsessed with baseball in a very un-Canadian way, and he cooks a mean latke.

I know a lot about Matthew, more through the time we spend together than by the things he says. And now that he lives thousands of miles away, I realize this is a rare and beautiful thing.

So, it was fitting that it was Matthew, upon learning of my stirring desire to slow down, who recommended I read *Better Off.*

I say fitting because of the type of friendship Matthew and I had fostered. The kind that took account of ideas before news. The kind that spilled over long meals and slow meanders along the Pacific Coast. Fitting because Matthew grew up in an old family cabin sheltered on a hill, in a bay, on a small gulf island off British Columbia's coast — the slowest place I can imagine — and, notably, the utter opposite of my suburban childhood.

I devoured *Better Off.* And soon after, I gave up the Internet.

All That Deadens, All That Dulls

We are part of a culture that can't bear to let our smartphones out of sight for even a few minutes. In fact, a recent online poll revealed that 24 hours a day, half of smartphone users keep their devices with them, within three meters to be exact. And when people say they take their phone with them everywhere, they mean *everywhere* — the bed, the boardroom and the bathroom.

And, with the help of these devices, we are living in a kind of medicated stupor, muted gray: happy, but not too happy, sad, but not too sad.

Katniss Everdeen, the heroine of Suzanne Collins's *The Hunger Games* series, finds herself bedridden in a sterile room hooked to a drip: "Morphling dulls the extremes of all emotions, so instead of a stab of sorrow, I merely feel emptiness. A hollow of dead brush where flowers used to bloom." (*Mockingjay,* p 218)

We know this feeling. When the extremes of real life encroach, we turn to morphling — our iPhones, our laptops — hitting "mute" on a moment.

My own desire to give up the Internet for a time originated from a growing restlessness and distractedness I recognized in my own life. I felt frustrated by the roundabout modes of communication that were developing in my closest relationships. A text from my mom: *Did you get my email?* An email from me: *Call me.* I felt like I was wasting reams of time, that most of my relationships had become a patchwork of online check-ins. I was motivated by the desire to discover the person and parent I could and would be offline.

Why was I in such a hurry? What was I gaining through my online check-ins? Why was I turning down in-person get-togethers? These were some of the questions I set out to explore.

I knew the lull of morphling. I knew it all too well.

> "We were all meant to be: spontaneous, free, aware, unafraid to love, without hubris: whole. Not as we are: fragmented, inhibited, sunside and darkside in collision instead of collaboration, so that we are afraid of all that we might find in the sinister world of the subconscious, are suspicious of intuition, and close our doors on the knocking of the Spirit."
>
> — Madeleine L'Engle, *The Irrational Season*

We Fast to Awaken

People fast for different reasons, but the fruit of the labor is universal: we are awoken — to our hunger, to our appetites and to the gaps in our lives. I abstained from using the Internet for 31 days in what I called my *Letters from a Luddite* experiment. When Internet columnist Paul Miller gave it up for an entire year and then returned to the "real world," he had this to say: "It hurts to be alive." Indeed, it does. And so it makes sense that we would want to numb.

Where we once used substances to ease our aches, to distract from the pain and mess of life, we now use gadgets. And like smokers huddled outside church after a long service, we've given each other permission.

> Sixty seven percent of cell owners find themselves checking their phone for messages, alerts, or calls — even when they don't notice their phone ringing or vibrating.
>
> — Pew Research

But what are we gaining from our attachment to the digital world? "We are not going to solve the technological problem with more technology," insists Read Schuchardt. "We lose our humanity when we submit entirely to technological systems. In order to regain our humanity, we must resist it."

Today's multitasking tools really do make it harder than before to stay focused on a single task such as long acts of reading and contemplation. According to one study, internet addiction changes the brain in ways that are similar to cocaine's effects.

> "The Word of God is very near to you, it is in your mouth and in your heart for your observance. See, today I set before you life and prosperity, death and disaster . . . Choose life."
>
> — Deuteronomy 30:14, 15, 19

In order to stifle the gray, we must break from our addictions to detox. Removing ourselves from the drip is the only way.

We might take a page from TV host George Stroumboulopoulos when he says in an interview with *Toronto Life* magazine: "I don't have a balanced life, and I don't want one . . . I need to be driven by high highs and low lows. I want emotional range in my life. I don't want my life to be easy. You know people say, I just want a good life. I don't fuckin' want a good life. What is that? I want to feel things."

We fast from the Internet to embrace the mess, to savor the highs and weather the lows. We fast to do something hard, to enliven our wills. To WAKE UP. We fast to find our way back to our humanness.

We Fast to Hear

> "It doesn't matter to me who's Prime Minister or who's sleeping with whom," shouts Sherlock Holmes, played by Benedict Cumberbatch in a memorable scene on BBC's *Sherlock*. "It's not important! Listen, this [pointing to brain] is my hard drive, and it only makes sense to put things in there that are useful — REALLY useful. Ordinary people fill their heads with all kinds of rubbish, and that makes it hard to get at the stuff that really matters. Do you see?" (Season 1, Episode 3.)

We are all publishers on the Internet. We think our innermost thoughts are profound, important and shareable. And, often, they are. But what they are not, is unique. If the Internet has anything to teach us, it's probably this: everyone's a star and everyone's a critic, and we might all do well to take pause before posting.

Baratunde Thurston, perhaps the most famous recent "disconnectionist," wrote in his *Fast Company* cover story, "#Unplug":

> "My friends say I'm the most connected man in the world. And in 2012, I lived like a man running for president of the United States, planet Earth, and the Internet all at once. Physically, mentally, digitally, I refused to stay still...
>
> Here's a partial quantification of the year: Cities visited: 34... Facebook posts: 1,518 (four a day). SMS threads: 3,702 (10 a day). Photos taken: 4,845 (13 a day). Tweets: 11,541 (32 a day). Gmail conversations: 59,409 (163 per day). Miles flown: At least 128,000, which is more than enough ecological cost to outweigh the benefit of my reusable shopping bag. By November, I'd reached rock bottom. I was burned out. Fried. Done. Toast... I considered fleeing to a remote island for a few weeks, but I realized I wasn't craving physical escape. I didn't actually want to be alone. I just wanted to be mentally free of obligations, most of which asserted themselves in some digital fashion. I decided to stay still... and step back from digital interaction. Yes, me. The recipient of the 2011 Shorty Award for Foursquare Mayor of the Year."

As Thurston slowed, he experienced "an expansion of sensations and ideas." Projects that had stumped him had new possibilities, conversations deepened, urgency waned. "The end," he wrote, "arrived too soon... the Internet and the world seemed to have gotten used to life without me."

Stepping offline reminds us that we are small. The online world, and indeed the world at large, keeps on without us. Our likes, our comments, and tweets are not missed. The world — it keeps on turning.

Baratunde Thurston challenged the world with these words: "I left the Internet for 25 days, and so should you." During that time, he made four major realizations about his life:

1. I had become obsessed with The Information.

2. I shared too much.
3. I was addicted to myself.
4. I forsook the benefits of downtime.

> "The greatest gift I gave myself was a restored appreciation for disengagement, silence, and emptiness. I don't need to fill every time slot with an appointment, and I don't need to fill every mental opening with stimulus."
> — Baratunde Thurston

Do you have an inner life, a belief in a higher power? One of the effects of our current age is difficulty perceiving and relating to God, and therefore our fellow human. Has your spiritual life deepened since signing up for Twitter? In the quiet of the morning, do you begin with meditation or, instead, with a quick scan through your email?

"Many voices ask for our attention," says Catholic writer Henri Nouwen. "There is a voice that says, 'Prove that you are a good person.' Another voice says, 'You'd better be ashamed of yourself.' There also is a voice that says, 'Nobody really cares about you,' and one that says, 'Be sure to become successful, popular, and powerful.' But underneath all these often noisy voices is a still, small voice that says, 'You are my Beloved, my favor rests on you.' That's the voice we need most of all to hear. To hear that voice, however, requires special effort; it requires solitude, silence, and a strong determination to listen."

> FOMO, the acronym meaning "the fear of missing out" cropped up a few years ago, but today a new buzzword in gaining traction: JOMO, the "joy of missing out." What joy it is to miss a weekend's worth of Facebook updates, to wave goodbye to a tidal wave of tweets. We fast to keep from drowning in a sea of voices. We fast to hear.

When we deprive ourselves of our digital technologies with the intention of making room for quiet reflection and stillness, we help develop self-discipline and fortitude, fostering a greater openness to God or whatever is sacred.

Cartoonist Michael Leunig's wonderful collection of prayers centers on an unusual image of a man kneeling in front of a duck. In the introduction to his book *When I Talk to You,* he explains the image this way: "But how do we search for our soul, our god, our inner voice? How do we find this treasure hidden in our life? How do we connect to this transforming and healing power? It seems as difficult as talking to a bird. How indeed?

"There are many ways, all of them involving great struggle, and each person must find his or her own way. The search and the relationship is a lifetime's work and there is much help available, but an important, perhaps essential part of this process seems to involve ongoing, humble acknowledgment of the soul's existence and integrity. Not just an intellectual recognition but also a ritualistic, perhaps poetic, gesture of acknowledgement."

By tuning out the noise, we open up space for silence and solitude. In quiet we better hear our deeper self, the place from which we can trust ourselves to make big life decisions and smaller daily choices that direct our lives. In fasting, we surrender our distractions to seek this silence and, in it, know the truth.

"Silence of the heart is necessary so you can hear God everywhere — in the closing of the door, in the person who needs you, in the birds that sing, in the flowers, in the animals," said Mother Teresa.

Engaging in spiritual reading and reflection remains a primary desire for most of us, yet we are busy. With our minds cluttered, it is difficult to hear. We must tune out in order to tune in.

We Fast to See

During Ramadan, Muslims across the globe fast from water from sun up to sundown. When night falls, they give thanks for this precious gift. In the same way, we choose to fast in order to see these everyday gifts, to renew our gratefulness.

"Practicing mindfulness, being open to God's voice, is made difficult by the "time anxiety" we so often feel," says Cecile Andrews. But slowing down gives us time to notice things: the taste of food, the smell of the air, the stranger passing you a smile on your walk home from work. Noticing these everyday gifts cultivates gratitude rather than feeding the discontentment we so often feel.

The Internet is an unprecedented tool, yet we expect it to give at all times in all places, even things it cannot give. When we draw away and then return to the Web, we see it again for the purposes it serves.

> "We're starting to see people yearning to be less connected and trying to implement rules, structure, and discipline in both their own and their families' lives, to ensure that all this connectivity does not come at the expense of relationships, skill development, and manners."
>
> — Randi Zuckerberg

We Fast to Go Deeper

> "The computer screen bulldozes our doubts with its bounties and conveniences. It is so much our servant that it would seem churlish to notice that it is also our master."
>
> — Nicholas Carr

> Are we indeed the first and only generation that has a deeper knowledge of sport or the latest sitcom than our personal and national histories?

"It is useful to think of the news as an environment," say Kevin Barnhurst and John Nerone, authors of *The Form of News*. "The newspaper industry, and now all Internet media, set up a panorama of distant events . . . and readers feel empowered because otherwise inaccessible places come within their reach. Against that backdrop, the newspaper sets up an intimate diorama filled with familiar faces and voices." When reading the news in print or online, they say, people get "jostled and annoyed, but feel smarter and better connected, if only because they know what to grumble about."

Think of the way you feel when reading, watching, or listening to your favorite media personality: we are drawn there because it feels both comfortable *and* unpredictable. Like the best kind of relationship, it is both familiar and exciting, and we keep coming back for more.

Social media and news sites, where we spend the lion's share of our time online, are designed in every way to keep us there for as long as possible. Because I am a freelance writer I know that online editors are looking for one thing and one thing alone: "clickable content." More clicks result in more pages and more ads seen. Simple. The Internet is monetized at every turn. Facebook remained ad-free, coasting on its cool cred until it had over a billion users, then they sold out. Make no mistake, Silicon Valley wants you online, and they want you to stay there. But the longer we remain online, the shallower our real-life connections can become. Our attention's diverted; we are spread too thin.

If we remain fixed to our screens, we rarely dive deeper, push ourselves to swim out farther than the familiar routes we travel online. Even ideas that strike us profoundly online rarely elicit more than a "like" or "share." If we want all of this knowledge to go further, our engagements to go deeper, then we need to take them offline. That's what good friends and good conversation do. They let us step out of our algorithm.

"One of the great challenges of today's digital thinking tools is knowing when *not* to use them," says Clive Thompson in his book *Smarter Than You Think,* "and when to rely on the powers of older and slower technologies, like paper and books."

We Fast to Know

Today I live in Toronto, but the last 30 years were spent growing, schooling, loving and working in Vancouver. As a result, a vast majority of my relationships are mediated by technology and limited to what people tell me instead of what I see.

But it's strange how we tell people stuff on the Internet. We announce it. We throw bits out. We photoshop our faces, our houses, our thighs. We edit our words. We show people our best. We overshare. (What is oversharing anyway? As if being too vulnerable, in our compulsively brave-faced culture, is a problem.)

When we draw away from our engagement online, we refocus our gaze at the world in front of us. We see our relationships, our dependencies, near and far.

Before I began my fast, I knew a handful of people on our street. Having little kids and a front yard swing helps with that. Though I had relationships with a number of neighbors, I relied on them little. We have most of the means to provide for our family and have access to services by phone and Internet. It's easier to Google for a pizza place than ask for a recommendation. More socially acceptable to buy a shovel than borrow one.

But one day, during my fast, when I accidentally locked my baby and my phone inside the house, I had no choice but to reach out for help. Scanning the street in utter panic, I spotted a car in a driveway a few houses down. After a few good rounds of knocks, Renee came to the door dressed in her housecoat. As soon as I'd blurted out my predicament, she ushered my daughter and me inside. She made a quick call to a locksmith, gave my kid a juice box, and then gave me what I needed above all: a hug.

> Raffi, the well-known singer and author, advocates online fasting: "When you're online and you're engaged, you're always kind of thinking about it like 'I wonder who's tweeting me? It's been an hour since I checked.' I would recommend it for anybody. Have a social media fast for a weekend or one day a week."

Twenty minutes later we were back in the house, the baby still asleep, and we were no worse for the wear. And while my daughter watched *Sesame Street,* I had my longest conversation with Renee to date, learning that she used to be a political speech writer before she began a second career in finance. My need drew me to my neighbor; she rose to the occasion, and through it we came to know each other.

We Fast to Remember

What is the thing in your life that you are most proud of? Is it the child you are raising? The marathon you completed? The project you hit out of the park? Think about how much desire, creativity, planning and work it required.

There are a couple of things that I am immeasurably proud of. One is the time I spent on my university rowing team. I remember perusing club tables in the quad, being drawn to a group of tall, boisterous girls and guys standing around the rowing club table. I took some information and stuffed it in my backpack. I liked the pictures of water, of the sleek aluminum boats slicing through the ocean at dawn. I graduated high school with the title of "female athlete of the year," having played four sports: basketball, volleyball, floor hockey and soccer. My friend Marieke and I had even made a surprisingly dominant force on the badminton court. As I read the rowing brochure, the competitor in me tingled with the thrill of taking on a new sport.

And then came the day of tryouts.

I knew no one on the team. I'd never been down to the boathouse. *Everyone will probably be better than me,* I thought. *Maybe I'll make a fool of myself. Maybe I shouldn't go.*

Reluctantly, I pulled on a pair of turquoise shorts and a long-sleeved cotton top and, at the last possible moment, sped off in my parents' maroon minivan. *I'll just go and scope things out,* I thought. *I can always bail at the last minute.*

I ambled up to the rowing compound where I could see a group forming around a strange-looking machine. An ergometer, I would later learn. People were taking turns pulling on it while Fahim, a small East Asian, cheered on and recorded stuff on a clipboard. He looked harmless enough, and I thought I could pull better than the last girl. So, I saddled up.

I was strong, tall and lightweight. Something like striking gold in that particular sport.

So I joined and started off as everyone does: bleary-eyed and frozen at 5:00 AM, one mere meter from the dock working on bobbing the oars up and down in the water. Balance exercises. Not quite the thrill of shooting a shell through ocean water that I'd imagined. But, slowly, our team managed to balance the boat, then we took some strokes, then we took longer ones, faster. Within months, we were clicking along watching the sunrise over the mountains. And before I knew it, our coach tried me out in the stroke seat — the lead rower in the boat. Month after month we trained five days a week, two hours on the water and two hours off. By the end of the season, we were killing it.

The work was beautiful. We were sleek, powerful, strong. I was an integral part of a team. There were eight seats and one coxswain. If one person neglected to show up to practice, the entire crew was benched. The sense of accomplishment was that much sweeter because it was shared.

Everything about our modern-day technologies, from the apps on our phone, to the gadgets in our cars, are centered on ease; but making things easier doesn't lead to a deep sense of satisfaction. Patience, discipline and hard work do.

I look back on the days I rowed in university with pride because waking up at 4:30 AM and pushing my body to the brink of exhaustion day after day took discipline.

I consider birthing my three children without the help of drugs or intervention a monumental achievement because it took Herculean work and extreme trust.

Difficult work leads to lasting joy.

The things we are most proud of require all of us. And once the ache, blood and tears are washed away, we are left with the solidity of our achievement.

Why Fast the Internet?

> "And here, for me, is another profound truth: understanding, as well as truth, comes not only from the intellect *but also from the body.* When we begin to listen to our bodies, we begin to listen to reality through our own experience; we begin to trust our intuition, our hearts . . . Truth flows from the earth. This is not to deny the fact that truth flows from teachers, from books, from tradition, from our ancestors, from religious faith. But the two must come together. Truth from the sky must be confirmed and strengthened in the truth from the earth."
>
> — Jean Vanier [Emphasis mine]

We are embodied beings. Since we communicate through body language long before we ever learn to talk, the language of gesture and eye contact is our original language. As we grow, we add words to our arsenal but language of the body remains our primary means of understanding and expression.

From birth, the eyes of humans by nature seek out the eyes of others. What happens when that gaze is broken so permanently that people don't know how to connect with each other anymore? What are the ramifications for empathy? For sexual intimacy? For parenthood? Too rarely we stop to consider: what is this new thing for? What is it replacing? Where will I find the resources or time to engage with it? How much is enough? We fast to begin asking these questions.

In a March 2014 article in *The New Yorker* titled "The Pointlessness of Unplugging," Casey Cep argues that "the unplugging movement, which encourages us to disconnect from technology, is unsustainable and misguided." Let us consider another proposition: The Internet-addiction movement, which encourages us to connect online at all times in all places, is unsustainable and misguided. Which is more accurate?

There are no simple answers here, to be sure. The terms "disconnecting," "unplugging," and "digital detox," are limiting because they suggest that offline/online worlds are separate when, increasingly, they are not. But these terms are a starting point as we personally and collectively navigate uncharted waters. Some people have a better time finding balance in their lives when it comes to using digital technologies, but others of us have a harder time striking a balance. Taking a break, carving out space without a screen, helps us lift our eyes, to engage in different ways that, for many, is life-giving.

Washington University psychiatrist C. Robert Cloninger says: "You're going to be confronted with many stimuli designed to grab your attention and addict you, so you should be very careful about what you expose yourself and your children to. Everyone needs time for quiet, reflection and reverie. You have to achieve an equilibrium between those deep human needs and other stimuli, or you'll have problems."

Suggesting that to unplug for a few days is pointless because people plan to return to the digital world is misguided. By that logic, we should also rid ourselves of weekends, summer vacations, fasting for religious or health purposes and giving up drugs.

> "Whether it is imposed by circumstances or chosen through spiritual discipline, [fasting] is about nurturing a posture that holds all things lightly, that ensures that our passions are subject to us and not the other way around."
>
> — Gregory Brent Pennoyer, *God For Us*

Let's return to our story of the two friends at the Wicker Park coffee shop. One felt the urgent need to send an email, halting an hour's conversation between friends.

We — you and I — have traded communing, being with people in the real world, for a counterfeit. But we can shift things simply. We can set our smartphones aside and set limits on our technology use — and this discipline can lead to our joy.

We must remind ourselves and reveal to our children the wonder found only in the real world: the beauty and peace in the stillness of the outdoors; the sense of accomplishment and joy in writing a letter to a loved one; the sheer delight of sitting around and playing a ridiculous board game. These are the things that will motivate us to set aside our smartphones, close the lids on our laptops and pay attention to the world in front of us.

And why is it imperative that we model and teach this to the young people in our lives? Because we can remember a time before the smartphone; they can't. We can and must be deliberate in our consumption now, in our workplaces, in our friendships and in our homes. Present people, people who look you straight in the eye, are a rare and wonderful breed. Imagine being that kind of executive, that kind of parent, that kind of teacher, that kind of friend.

It is within our grasp.

What Sustains Me Most Deeply

Esther Emery is the youngest daughter of the back-to-the-land educator Carla Emery. She chose to fast, giving up the Internet for 12 months, long enough, she says, for her "emotions to morph from constant performance anxiety through disdain and detachment back into an honest desire to connect." After disengaging from the Web for a year, she returned with a flourish and now blogs from a yurt:

"Maybe [what you need in your life is] radical self care," she writes. "Maybe it's a choice to do a little less of whatever makes you feel empty inside, even if it's all supposed to be super fun and awesome! Maybe it's a purchasing blackout, or making a bonfire out of all the magazines and catalogues that promote illusions of wealth. Maybe it's skipping the party that makes you feel like crap. I invite you . . . to do a little more of what it is that heals you. Whether it is community and ritual and togetherness. Or silence and snow-covered trees and sweet warm drinks."

Some people call it a fast, others a resolution, some a detox. Whatever you call it, it's about abstaining from what deadens and dulls to seek what calls you awake. It's about leaning into what sustains you most deeply.

What sustains me? Trail running, rowing, peaceful moments with my children, good conversation, ferocious laughter with my ragtag collection of brothers and sisters, listening to Cat Stevens, writing poetry, Saturday adventures with Michael, exploring New York alone. And flying kites.

Chapter Challenge:

We fast to draw close to this question: What in the world moves and sustains me most deeply? Write your answer here:

____

____

____

____

This is not about "don't;" it is about "do." It is about paying attention to what enlivens us and feeding that hunger. It is, as Albert Borgmann says, "creating the positive conditions in which technology becomes less compelling and different engagements thrive and flourish."

It is about being fully human in a smartphone world.

7

Gaining the Time

Implementing Constraints

A dead thing goes with the stream.
Only a living thing can go against it.

— G. K. Chesterton

ALEXANDRA LEIKERMOSER, owner of the yoga studio of Toronto's Ritz-Carlton, enforces a no-phones-at-your-mat rule. Her clients, primarily high-powered Bay Street executives, take issue with this. "Are you a doctor on-call?" she politely inquires. When she gets the expected ring of "no's," the ladies reluctantly tuck their BlackBerrys in their Lululemon bags only to dash out as fast as they can say "Namaste," once class is finished. Forty-five minutes is simply too long to be away.

We can sense that there is something wrong with our relationship with time. For most of us, "we're rarely aware of what we are doing," writes Cecile Andrews in the wonderful compilation *Simpler Living, Compassionate Life*. "Our attention is constantly diverted. Being mindful is difficult because we are always anxious about time. We never have enough of it."

One man claims to have found a solution. In *Better Off*, Eric Brende makes the persuasive case that most of us would have more time and enjoy that time more by radically minimizing our reliance on modern technology.

The notion that technophobes are backward gets turned on its head as Brende and his wife realize that the crucial technological decisions of their adopted Amish-like community are made more soberly and deliberately than in the surrounding culture, and the result is greater — not lesser — mastery over the conditions of human existence.

They found that a life free of technological excess can shrink stress and expand happiness, health and leisure.

In our accelerated culture, we complain about having no time, all of the time, and yet we impulsively spend what free moments we have submerged in the never-ending drama of email inboxes, social media feeds and reality television shows "that leave us no more enriched, no further ahead, than if we'd just not bothered with them in the first place," writes Jayar La Fontaine of Toronto's Idea Couture. "What's worse is we've valorized this ability to keep a more and more frenetic pace on the data treadmill."

For all the affordances of our new communication technology, we haven't established healthy norms or habits as a culture and it's costing us our time, creativity, energy and relationships.

It's a treadmill of our own choosing. We can step off.

> "I had this book I was meaning to read and, when the power outage in Toronto happened, I read a novel in one day. I had nothing to distract me. I wrapped myself in a blanket and just read. When you actually give yourself time, you can do so much."
>
> — Krittika Sharma, textile artist

Going Offline

A couple of years ago I began feeling uncomfortable with the demands of the Internet on my life and the world at large. While the benefits of the Internet are numerous — Skype and photo sharing, for example (I have seven brothers and sisters scattered around the globe, so I know how awesome these tools can be) — they are mixed with an unrelenting persistence for our attention, and it is this I found unsettling.

It was around that time that I came across a documentary in which there was a scene of a priest conducting a blessing service for smartphones. Here was a man, dressed in holy vestments, calling on the God of the Universe to bless a BlackBerry. I had a visceral reaction — complete and utter shock — at the scene.

Blessing blackberries — actual harvested food — to our bodies' use would have been one thing. But blessing a small 2.3 x 4.5 inch piece of glass, metal and plastic designed to speed our lives to the brink of exhaustion, is quite another.

There was something wrong with this picture.

The centrality of Internet technology in our daily lives made me squeamish, and I felt I needed to figure out why. I had suspicions.

I had grown tired of Facebook mediating my relationships. Like missing a friend visiting my city because she'd posted to my "wall," and I didn't see it. I felt the Internet made me lazy as a thinker, a writer and a friend. I knew it enabled me to emotionally disengage, as I whiled away my time with mundane, ever-ready filler: contextless information via newsfeeds, Twitter and the like.

The truth is, I was both bored and obsessed with the Web.

I stepped offline for 31 days and chronicled the journey by typewriting and mailing a letter to a friend every day. No Google mapping, no email, no Facebook or online news. I was searching for life beyond our steady state of distracted connectedness. Setting the intention to write 31 letters got my hands moving; it disciplined me to write every single day, with or without kids clambering about my knees.

Following are some of those letters, and what I discovered along the way...

Jan. I, 20I2

Happy New Year, friend.

We ended up on a last minute trip to Ottawa to ri ring in 20I2 and got home a few hours ago. Today is day one of this internet fast and my first feeling has been a sense of RELIEF.

I don't have to check email. I didn't have to say yes to a video on netflix when we got home. Instead, Madeleine helped me with the dishes (aka 'played bubbles') and we had a tea party.

I realized today that I write notes/reminders to myself by email on my iPhone, so ~~I8V-~~ I've gpt to put a notebook and a pen in my purse to jot things down. Another hiccup popped up today when Michael and I were talking about finally getting around to buying a filing cabinet. He suggested Craigslist which, of course, I can't use this month. So that will fall to him and I'll check Ss Staples in person tomorrow.

I am going to be writing these letters to you with all of the minutiae of the day, but tonight all I've got is a quick update because my eyes are falling shut...

That's all for today, sweetffiend.
friend.

Until tomorrow.

xo
Christina

January 2, 2012

Dear Marisa,

Before I started this fast I jotted something do wn in my journal:

"I think I may stop worrying about what other people are doing... and just do my thing."

The Internet allows for unprecedented comparison. People showing us glimpses of their lives, often (always?) just the best of them -- accomplishments, best photos, best days -- I know I've fallen prey to this.

Today I found myself wondering what's happening online: who's getting my automatic email message, what's happening in my friends' lives updated via facebook. I even thought about a slight tweak I wanted to make on the Letters from a Luddite blog's original post - but I can't... I haven't yet experienced this as a frustration. It's actually felt freeing to not make the correction.

Today was a sick day in our house. Michael mpped around the house draped in our Hudson's Bay blanket and I was with the kids. Madeleine helped me grocery shop - collecting vegetables and fruits for our 'new year diet' - no wheat, no sugar.

I'm finding all kinds of fun things to send you. I must have a collection of more than 200 postcards, so expect some of those coming your way.

Today was the first day I pulled out the Yellow Pages rather than do a Google search. I needed the phone number for the local community centre to find out leisure swim times. I saved the number in my phone. I think this fast may make me more efficient in this way - having systems for saving information/ important contact info rather than looking it up over and over. Do you do that? I don't know when it started. I used to have an amazing memory for phone numbers and that went straight out the window when I got a cell phone and I didn't need to remember anymore. Hmmmm...

Happy Monday, friend. Still no snow here. Hoping for some sledding weather soon.

Love,
Christina

Jan. 6&7, 2012

" For tribal man space was
the uncontrollable mystery.
For technological man
it is time that occupies
the same role. "

- Marshall McLuhan
The Mechanical Bride

Dear Marida,

I am noticing a lot of my habits.

Like today, at church, I got the phone number of an acquaintance named Nick. I've met him a couple of times at Grace Centre for the Arts events and I got his number to borrow a book for an article I'm writing about a mutual friend of ours. I know Nick is a poet and today I keyed his last name in my phone. Later today I found myself wondering about his work and where he's published and, naturally, I wanted to Google him. But I couldn't. I realized that I Google people all of the time -- particularly people I first meet, acquaintances I find intriguing, names I come across in books, online, in the paper. But I know Nick and I can't Google him so next week I'm going to Google him face-to-face, get the answers to my questions first-hand.

Habit #2: Checking in with most friends via facebook/email... "Hi, thinking of you..." Scroll through their recent posts, see how things are going. But is this really checking? Am I getting the real picture? Most of us don't pour out our needs/challenges, big or, small, on twitter and the like. In the last week I've had two long phone conversations with best friends in Vancouver. Through them I learned so much more than the sum of their online updates. Full disclosure: they initiated both calls. Thank you, friends.

(1)

(Nervous) Habit #3: Fiddling wi th my iPhone WHENEVER. I must check my email an average of I5x a day, if not more. In addition to that I check my facebook newsfeed a couple of times a day and at least one of my websites once a day (etsy, blog, seekingeve, etc.) I check email a lot while I feed the baby because my phone is always on hand. I've since replaced my phone with a novel. These little online 'check-ins8 add up to a lot of time. this past week alone I finished a(n amazing) novel, read a paper start to finish (only parents will know what a miracle this is,) and begun a couple of other titles ('only as good as your word' by susan shapiro, and a book on the history of Canada called 'forming of a nation.8) Last night Michael andI lay in bed reading bits of the paper, passing pages back and forth and chatting like we used to before the kids arrived. The clip I'm including is from last week's Globe. (This weekend's edition got soaked on our front porch. Totally unreadable. Score one for the Internet: Globe and Mail.ca does not get water-logged.)

Habit #4: Turning to Safari before turning to people for help. On Thursday, when I locked my phone (and my baby) in the house I had no choice but to find a person to help. Left to my own devices I would have been happy to rely on my handheld device to get me out of the bind. What my iPhone wouldn't have given me, however, was a hug (which I needed above all), a juice box for my kid, a warm place to wait for a locksmith, and conversation with my neighbour, During those tumultuous 23 minutes I learned that Renee used to work as a speech writer on Parliament Hill. For six years she wrote for senators before heading off to an MBA and sta rt a second career as a broker for a major bank. I had no idea. Had I not been stuck without my smartphone in hand, I would likely neve have struck up this conversation. Grateful for this neighbourly care, Madeleine and I brought over a batch of cookies wrapped with ribbon.

Adieu good friend,

Christina

January I2, 2012

Dear Marisa,

Today you get pretty paper. I haven't written any of these letters on the computer because there is the tendency to edit and then I overthink what I'm trying to say, so I find the pen or typewriter best.

There is no question that the Internet is an incredible tool. I had hoped to file two stories before the fast began but I didn't finish them (in one case, start) before Jan I. Both were due January I5. So, in my true deadline-loving nature, I finished them with little time to spare. But I had to spare extra time because I couldn't email the files. They had to be couriered by USB.

So, this morning I woke up at 6am to write my last 500 words, make some edits and save them to 2 USB sticks. I wrote letters to the editors and prepared the envelopes last night, planning to dash off to Canada Post before Michael's work day began (he was working from home today.) I got up and discovered our first real snow -- a full-blown blizzard -- dumping down outside. I finished writing in record time, desnowed the car and began the snail's pace drive to the post office. I traveled IO blocks in 20 minutes. Once inside, I found out it was going to cost me $36.00 (exclamation POINT) to send ONE by Express Post over the weekend. All in it was $50 to file both stories when I could have done it for FREE over email.

Lesson of the day: filing stories by email is AWESOME.

(I2 cont.)

Aside: I also made Michael late for work.

A few other things about trying to interview/ write articles without the Internet:

I. Peope want to give you information over email rather than explaining everything to you over the phone (which is totally valid.) Also, receiving typed out information makes it impossible to misquote someone.

2. Fact-checking online is a breeze.
c

3. Tracking down quotes from books or articles peope reference is difficult without the

Internet, but also an opportunity to source material in different ways. One interviewee, Alyssa, referred to a book by Richard Wurmbrand. Having difficult locating the title at the library, she referred me to her friend who had a loaner copy. He delivered it to my house personally, and I've been enjoying reading it in its entirety: something I wouldn't have done if I'd just grabbed a quote online.

How's the new job going?

Love you.

Christina

January I4, 2012

Dear Marisa,

So, a couple of bumps, to say the least. My editor at a BC-based magazine just called to say that she hasn't received my package. The one that cost me a whopping $36 ver never showed. When I went to the post office today they could tell me it was 'processed' but confirmed it was 3 days late. They pointed me to a toll-free number on my tracking receipt and told me I had to call for my refund, which I did...later in the day while juggling my two kids. the young operator informed me that my 'package8 - a skinny standard envelope with a single page and USB stick - was in Richmond. She proceeded to explain that they can't process my refund until the package is delivered to its destination (North Vancouver) but who knows when that will be? The good news is I will be getting that $36 back, the bad news is I had to have Michael email the editor the piece last night because it was already days overdue. Boo to Canada Post.
...(sorry guys, I love you, but you seriously dropped the ball on this one...)

Whew. Glad to have that off my chest.

Yesterday I called my mom looking for the Vancouver Sun. I wanted t eir address. I called her again today looking for the phone number of a travel agent friend of ours (I am thinking of meeting up with my old roomie, Sarah, for her 30th birthday trip to Vegas) but she didn't have it.

Her exact words: "I EMAIL her. I'm not 9-I-I, you know,"

Apparently she isn'te enjoying the 'personal

touch' of the phone as much as I am. And I think she meant 4*I-I.

So nice to hear your voice today. I hope the letters make it to you. So weird they are coming in such a random order.

xo. Chris tina

Dear Marisa,

People keep asking me how (oops, sorry, i hit something with the typewriter) how the fast is going. I am truly loving it. There have been some frustrations: lost mail (the $36 package is now 'under investigation'), not being able to check or post things on Craigslist (seeking a filing cabinet and more scout badges,) not being able to grab the laptop and call the grand-parents for a quick Skype date when the kids are doing something ultra cute. But the good far outweighs the bad.

I am reading. My head is clear. I feel freed up to not be resposible for knowing every banal thing going on in my friends' worlds via facebook. I thought I would be hooped without Google maps on my iPhone but I've remembered my way to places and asked for directions over the phone. I am getting mail too.

Yesterday two letters came in the mail. One was a gorgeous envelope adorned with a dried flower and lovely handwriting, the other with brightly coloured circles, each with a letter spelling: F R I E N D. Laura told me about a trip she took to Fanny Bay and how the sea lions kept her awake at night. Julia penned many pages, telling me about life, faith, marriage, love. Nothing compares to the hand-written note. Madeleine and I check the mailbox every day, carrying in treasures each week. A letter. A package from Grandma. An envelope from her Auntie Brittany filled with pictures and a description on the back of each. We sit cross-legged on the floor and carefully open each one, pieces of ourloved ones traipsing off the page and into our daily life. Letters are tangible, especially to Madeleine, in ways emails are not. She understands that someone sat down and packaged that card or sweater just for her.

One of the things I am tucking in with this letter is an article

about the return of the abacus. According to the Globe and Mail, the ancient bead-based calculator is making a comeback in after school programs across Canada. I am not no mathematician but I get how something so hands on would help learning. I remember in grade 12 math, we were strongly encouraged to get one of those $100 algebra calculators. I pushed back and said I didn't want it, hoping to learn the long-hand process to the answer. It took more time but, for a slow learner like me, it finally made the numbers click and I could see why I was arriving at the answers on the page rather than just copying the figure down.

That reminds me, when I was 16 I did something similar at the

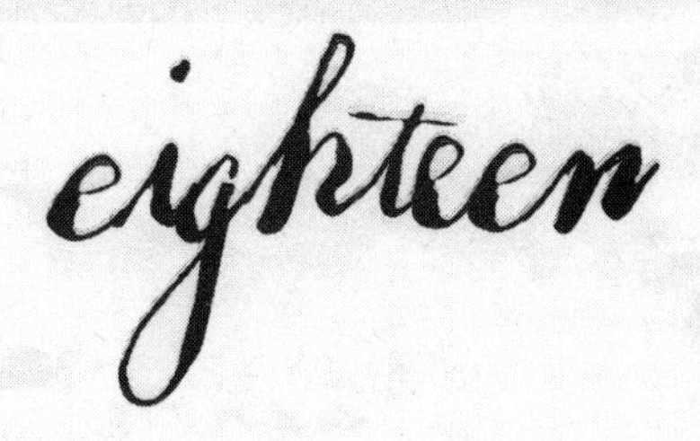

/...2

(2)

bank. I went in to do a deposit and get my bankbook updated when the teller informed me that I had to get a bankcard, a new ATM one. I told her I didn't want it, explaining that I didn't want it to be easier to spend or take out money. She told me the whole system was moving over and I had no choice. At the age of I6 I had the balls to ask for the manager (I get it from my mom's side.) The manager confirmed what the teller said: I had to get this new card but, he explained, I could use it just at the bank, with the teller, to confirm my account and proceed the way I always had. "Leave it an envelope at home when you're not using it," he suggested. We know how that story goes. Banks want us to spend more money, bank more fees. The card went in my wallet and my saving was never the same. I am not trying to come off as this weird anti-technology kook, but there is something about the tangible -- touching, handing, exchanging -- ~~exchange~~ of things that makes them real. The advent of debit caused spending rates to ~~increase at higher and higher rates.~~ sky rocket.

I am still mad that I had no choice in the matter. But there are lots of things we do have a choice in, especially with our kids. I want to teach Thomas and Madeleine about money, I want them to play and learn from things in their hands, not just on the screen.

William, owner of the Outhouse coffee shop near my house, showed me a demo for a new iPad/~~Tablet~~-/tablet product for kids. It's called augmented reality, developed for Sesame Street workshop by Qualicom. Children take Sesame Street chracters/figurines and move them around on a special mat. Then they hold the tablet in front and the software plays out a scene involving the chracters. Is this engaging kids more than simply sitting and watching Sesame Street, like my husband suggests, or is it yet another way we are eliminating an opportunity for imaginative play? What do you think?

Hope this wasn't too heavy.

Love to you, friend.
The snow is slowly drifting outside our window.....

xo Christina

Marisa, (exclamation point)

It's possible to get reacquainted with old schoolmates without Facebook (exclamation point x10)

This morning I drove through blistery snowy weather to meet up with my aunt Astrid in Leslieville. I double-took on a woman with pulled back dirty blonde hair almost immediately. Though I misremembered her name, it was in fact who I thought it was: Eva Danielson, from my middle school. She was a year younger than me in school, with wild, gorgeous curls and daughter of the music teacher. It turns out she has lived in Toronto for five years, has 2 kids (Ella, 4 and Rosalie, 1), teaches music and her husband works for the Luminato arts festival. Amazing. We are going to try get together again at the same cafe where she was leading a musical circle time for toddlers.

It's a small world.

The wonder of this encounter got me thinking about a key piece in this off-line experiment : TRUST.

Trust that moments like this will happen. It's magical. Neither of us had a hand in it, but Providence -- God -- did.

Online there is always someone initiating, but here, in the 'real world', we are left to these chance, or not-so-chance, encounters. It's a wonderful feeling. Love C.

today is a blue-grey kind of day.

i am missing my friends. missing you. missing sitting down and spilling secrets over really good tea. missing having you and avital and hoda and meg a short drive away.

but the internet won't help me today. because what i miss doesnt't fit through cables, doesn't travel well along the information highway.

i can't squeeze my children's enormous eyes, my questions about the future, the look on my face when you remind me about that time when you got rained on when we were roommates and i just had to have the window open, i just can't fit that on the screen.

only two more weeks until i am in vancouver.

love christina

After 31 days, how was I feeling, what did I find?

Relief

When we step offline, it's amazing how little we miss. It taught me how to trust that the world keeps on turning without my words, without my likes and dislikes. It made me feel small. It showed me I am not as important as I think.

Perspective

We are little gods on the Internet, crafting and controlling our image on the World Wide Web. My sister and I joke that we should start a site called real-mom.com where we do weekly photo challenges like: "Go take a picture of your toilet seat. RIGHT NOW." It wouldn't be pretty. And much of life isn't. But it is real.

Closeness

There is something about the therapeutic clickity-clack of the typewriter that allows for a different kind of writing. The kind that spills from the heart rather than the head. The kind that's intended for a single, known reader rather than a large, unknown audience.

New Habits

We are what we repeatedly do. I learned that the smartphone check-ins I make multiple times a day are not actual time-savers but time-suckers. That if I, as a mother-of-three, want to engage with new ideas, read books, study, create — then I have to save up all of those two-minute, one-minute, ten-minute windows and bank them for things I really want to do. Like write poetry. Phone my grandma. Read *The New Yorker*. Get things done. Build new habits.

Beauty

My online fast revealed the beauty of unplanned moments, reminding me that chance encounters beat out an online connection any day.

Quiet

I discovered a peace, the quietness of mind, I had been hungering for.

I also learned that in a world that won't stop talking, it is snail mail that gets people's attention.

Seek and You Will Find

I began my digital detox seeking to:

- enliven my real relationships and filter out the surface ones
- remove my go-to time-fillers: newsfeeds, email and Facebook
- challenge myself to engage with the new ideas, books, conversations I'd tend to otherwise miss
- open my ears to God's voice and my eyes wide to the world around me, while the hum of my online life fell quiet

Chapter Challenge:

What are you seeking?

Peace? Time? Intimacy? Something more practical? Take a few minutes to sit with this question, then record your answers here.

____

____

____

____

8

Quitting the Comparison Game

Reclaiming Delight

The purpose of education with us, like the purpose of society with us, has been, and is, to get away from the small farm — indeed, from the small everything. The purpose of education has been to prepare people to "take their places" in an industrial society, the assumption being that all small economic units are obsolete. And the superstition of education assumes that this place in society is "up." "Up" is the direction from small to big.... The popular *aim of education is to put everybody "on top." Well, I think I hardly need to document the consequent pushing and trampling and kicking in the face.*

— Wendell Berry

What is wrong is a world wherein we constantly see others as opponents, where we are enjoined always to compete.

— Joshua Paetkau

I USED TO THINK I'd be the kind of woman who brought home the bacon and cooked it, too. I shared this ambition with millions of women, raised by second- and third-generation feminists clamoring to smash through glass ceilings. *Marry later!* echoed the call. *Do it now, before the children!* came the cheers. I said "yes" to every opportunity. Took interviews on

Parliament Hill. Pitched a show to MTV. And then I married at the respectable age of 27.

Soon came the children. One. Two. Three. The learning. The nursing. The laundry. The lists. And, while I've worked to keep bringing in the bacon, the truth is: I mostly cook it.

So, in the spirit of full disclosure, here it is: I am sitting in the egg chair in my kids' room typing in the dark. My daughter is rolling around in her bed singing a hybrid lullaby about tinker faeries. I should be grinning ear-to-ear at this adorableness, but I'm not. I'm tired. It's late. Our son is teething and my husband, Michael, is off on a work trip — the sixth in the last two months.

This is the first time I've ever carted my laptop into their room at bedtime. The lights are out, so here I am sending invoices in the dark.

Downstairs looks like a cloud descended on our house and rained down bits of yogurt, salmon, cheerios, sand and peas — INSIDE. The kitchen sinks are oozing dishes, and the countertops are buried beneath leftover mother's day cake, corncobs and bags of groceries.

I drove the wrong way down a one-way street this afternoon. A Starbucks-toting-brunette's wagging finger brought this to my attention. My son's car seat wasn't properly fastened. My daughter was still in her pajamas. We were out picking up a water play table someone had left on their curb.

My hair is overgrown. I cut it myself last Thursday. A desperate last measure before boarding a plane to Vancouver. The gray/whites I've inherited from my mother are beyond camouflage.

While magazines on newsstands may carry my byline, while I've got a TEDx talk under my belt, and a LinkedIn profile that reads like a serious professional, the truth is this: my children, my compost and my magical-never-ending-hamper are where my hands and heart are firmly set.

My real life looks nothing like my online profiles. Chances are yours don't either.

Getting Real

Before I began my digital detox, I jotted a thought in my journal: *Maybe, if I step offline, I'll stop worrying about what other people are doing and just do my thing.*

We are a competitive tribe, ever hungering for more: more prestige, more comfort, more money. In my reading on the topic, I have learned that the First Peoples held cooperation as the highest value and competition as the lowest. Aboriginal cultures also believed that people who took more than they needed, as the settlers claiming their land did, had a form of mental illness. It was unthinkable to leave weaker members of the group with little or nothing when the strongest had much to share. Competitive, compulsive consumption at the expense of others equals a mental illness? That's something to chew on.

Today the Internet enables unprecedented comparison, and comparison is the secret killer.

We show each other glimpses of our lives. Often, if not always, revealing only the best of us: our accomplishments, photoshopped images and thoughtfully crafted status updates. Through them we slowly distort ourselves and one another as we compare careers, homes, children and relationships through the tiny windows we post online. What we eat, do, wear, buy — and how we talk about it — is all part of how we see ourselves or, more importantly, how we want the world to see us, says Andrew Potter in his book *The Authenticity Hoax.* He calls it *conspicuous authenticity.*

What would our world be like without the conspicuous authenticity the Internet breeds?

"I know the yearning to believe that what I'm doing matters and how easy it is to confuse that with the drive to be extraordinary. I know how seductive it is to use the celebrity culture yardstick to measure the smallness of our lives. And I also understand how grandiosity, entitlement, and admiration-seeking feel like just the right balm to soothe the ache of being too ordinary and inadequate."

— Dr. Brené Brown, *Daring Greatly* (p 23)

"Humility is not thinking less of yourself, it's thinking of yourself less."

— C.S. Lewis

Trusting Our Own Ideas, Thoughts and Insights

I love hearing my friend Caroline Senger share stories about growing up in a small village in the Black Forest region of Germany. There weren't many other kids her age around, so she spent a lot of time alone in the woods making up stories and going on adventures.

"My childhood felt very magical," recounts Senger. "The streets were cobblestone, there were fruit trees in my backyard, and there were always chickens and cats to play with. I used to watch slugs leave their trails that sparkled in the sunlight and think that they were magic. I would sit in the cherry tree, eating the fruit, and have naps in the garden."

Every person Caroline met seemed like an interesting fictional character, a part in the novel unfolding in her imagination: the strangers by the fountain, the kind old men in town who'd tell her stories and make her honey sandwiches. Sometimes her mind would play tricks on her — like the day in the woods when she found a pile of wood chips. The closer she got to the center, the hotter and hotter it got, and the more convinced she became that it would light on fire and burn down her nearby village.

It wasn't until she was a teen that Caroline began putting her thoughts down in journals. "I wrote them because I was reading a lot," she says, "and had a lot of feelings about the world, but no one to really share them with. I was reading feminist literature and German fantasy books, and the girls in my class were into sports and the New Kids on the Block. So, I wrote love letters to imaginary people in my diaries."

The Internet wasn't yet a part of Senger's life. It wasn't until her late teens that she got an email address.

"I felt very different and unique because I didn't know anyone else like me," she explains. "I had a lot of anxiety, and poetry was a way of dealing with that. I took myself seriously. I showed some of them to my teachers, and they encouraged me to submit them to a high school writing competition where I won two years in a row. I felt confident about my voice and my writing. I even performed a few of my poems at an assembly in front of my school. I don't think the students were into it! But the teachers praised me, and that was all I cared about."

She began spending more time online. First on sites like MySpace, then Facebook, Pinterest, tumblr and people's blogs. Slowly, over time, her view of herself began to wane. She began to think herself far less interesting than she had once believed.

"In that time, I developed a strong inner critic," says Senger. "I compared myself to other people, other artists and writers. I stopped writing, I took myself less seriously. But I wish my inner voice wasn't so critical, that I could channel the uniqueness and inspiration of my youth. I really started to believe that there are many other people more interesting than I am, more talented, more creative. I became cynical.

Something my childhood self could never have imagined."

There was once a time when the only people we had to compare ourselves to were the people we could see in front of us. Our brother. Our cousin. Our friends. Today we can measure ourselves up to the 2.5 billion souls online.

There is one person in my life, a fellow writer, who, at one point, I had to intentionally unsubscribe from because the comparison — and ensuing jealousy I felt — were palpable. I found myself wishing for the demise of this rising star who, only a year earlier, had been neck-and-neck with me career-wise. I discussed this with an editor friend of mine who confessed to the same kinds of professional jealousies — jealousies that would be far more difficult to breed without nibbling on things online: professional websites, articles, blogs, Facebook, Twitter and Instagram updates. In the past, this information would have taken months, even years to gather; now it takes a matter of minutes.

Scarcity, the never-enough problem, lives at the heart of our jealousies. It is rooted in the belief that we are not enough, that we don't have enough: time, money, success, or whatever it might be. "The feeling of scarcity thrives in shame-prone cultures that are steeped in comparison and fractured by disengagement."

— Dr. Brené Brown, *Daring Greatly*, p 27

Bless Others in Their Success

My colleague shared with me how she'd finally nipped these types of jealousies in the bud. "Bless others in their success," she said. *What a radical idea.*

"As a recent college graduate, it was unhealthy for me to compare myself to my peers while looking for employment. A month before, I had also ended a relationship that was defined by my life in college. Instead of blocking my ex and continuing to ingest the fake facade of my peers lives and their 'successes,' I just left the building. I now am really happy with my new job and developing a new lifestyle. Leaving Facebook has helped me move on."

— man from Kentucky, via the *New York Times* blog

Can you imagine a world where, instead of comparing, we blessed one another in our successes and said sayonara to tracking each other's progress?

The definition of jealousy is envying someone for their achievements and advantages, while experiencing such negative emotions such as distrust, suspicion and anger. (Source: Oxford Dictionary) *Distrust, suspicion, anger* . . . More than anywhere else, we are feeding these emotions and attitudes online.

> Imagine a world where we bless one other in our success and say sayonara to tracking each other's progress.

Bless is the opposite of curse. It literally translates to "confer or invoke divine favor upon; ask God to look favorably on." In other words, blessing is not just wishing the best for someone else, it's asking the universe to give them even more.

Blessing others is like slapping jealousy in the face.

Nothing Is Ever As It Seems

Through well-crafted status updates, photoshopped images and lists of accomplishments, we traipse around online like whole human beings when, in fact, we are broken.

As a young girl, I dreamed of the life Karen Kain had. She was, at that time, the prima ballerina of my home country of Canada. My mother bought me a copy of her biography for my tenth birthday. I can still remember its glossy black cover with a slender woman bent low tying her pointe shoe. To my childhood mind, Kain's life was as flawless as her porcelain skin. I knew nothing of airbrushing or editing, only this picture of perfection.

Fast forward 20 years. When I see advertisements for the National Ballet I still envision dancers lounging in their stately mansions, sipping cab sav and looking effortlessly beautiful. That was until I met one on my block. Lisa and I both have young kids; our daughters are in the same kindergarten class. She lives in a little green bungalow with a sandbox out back. Her life is not perfect. I know this because we stand outside in the blistering cold and she tells me so.

> "[The Internet] makes us realise how much our everyday lives suck compared to most. #justsaying #imsoboring"
> — @Zaedum

I imagine Karen Kain's isn't so flawless either.

Everybody Is Just Like Me

There is power in remembering that everyone is just like me.

Me, with the messy house, the boring Saturday nights. Me, with the failed Pinterest projects and imperfect romances. Me, who inflates how exciting day-to-day life is, how wonderful Mr. Right is, how fabulous the garden looks, how well business is going.

Everybody is just like me . . .

Who's hiding:

____

Who thinks:

____

Who struggles with:

____

We think it is just us. We think we are the only ones. We think nobody knows. But the truth is: we see it, we know it, we do. We all have our closets full of secrets. We are all bound by the minutiae of life. Celebrities flush toilets. Presidents brandish scars.

Because online everything is edited, every post delayed, each word and image mediated, the Web has a way of making even the dullest things look shiny. We piece together the life our prettiest, most successful-looking neighbors lead. We imagine the perfect lives they live. Inside, however, they're makeup-less and fighting.

> "Those who have a strong sense of love and belonging have the courage to be imperfect."
>
> — Dr. Brené Brown

At one time, people literally aired their laundry outside — until a revolutionary new technology called a clothes dryer brought it in. *Proximity paints the real picture.* Even if we air our foibles and grievances — our metaphorical laundry — online, only those who live closest to us are privy to our day-in, day-out attitudes and experiences. Real life paints the whole picture; online, we only see the strokes.

Addicted to the Internet

Since 2000, Asia has led the global digital market, now with over one billion Internet users. (North America lags behind, with a quarter billion.) According to estimates, more than half a million of Japan's children ages 12 to 18 are addicted to the Internet. An excess of screen time has been linked to sleep disorders, a child's risk of depression, attention problems and obesity, as well as interfering with their school work. In fact, a research study has shown that for every additional hour kids spend online, their happiness decreases eight percent. (Source: Happify website, 2014)

Nicholas Carr addresses the issue of Internet dependence and addiction in his book *The Shallows:* "The Net delivers precisely the kind of sensory and cognitive stimuli — repetitive, intensive, interactive, addictive — that have been shown to result in strong and rapid alterations in brain circuits and functions" (p 14). In other words, the use of social media offers the same repetitive, addictive, dopamine-inducing "reward" and expectation that keeps users hooked.

Japan is the first country in the world to institute state-run "fasting" camps for Internet-addicted children. The Japanese Ministry of Education has asked the government to fund these detox programs designed to help kids unplug from the digital world and reconnect to the real world.

The Japanese detox camps are to be held outdoors and at public facilities where kids would have no way to access the Internet. Instead, they would be encouraged to participate in games, team sports and other activities. Psychiatrists and clinical psychotherapists would be on standby in case the youngsters struggle with the transition back to reality.

As many as a million young people in Japan are thought to be holed up in their homes, some for decades at a time, spending their waking moments immersed online, reports the BBC.

These people report that as time passes, a deep loneliness can set in and, with it, a deep sense of shame.

"Loneliness is a feeling of not being a part of anything, of being cut off," explains Jean Vanier, who spent decades working with intellectually disabled people through L'Arche communities around the world. "It is a feeling of being unworthy, of not being able to cope in the face of a universe that seems to work against us." It can feel like immense guilt, says Vanier. Of what? Of existing? We do not know. "Loneliness is a taste of death."

(Source: Jean Vanier, *Becoming Human*, p 33)

In Japan, there's a term for these home-bound Web users: *hikikomori*. Hikikomori are people who act as recluses, preferring to spend their days and nights mostly on the Internet, especially on Japan's 2ch (a popular textboard). The longer hikikomori remain cut off from community, the more aware they become of their social failure. Their self-esteem and confidence are drastically reduced, making the idea of returning to the social masses more terrifying the longer they stay away. Hundreds have taken their lives.

> "Connection is why we are here. We are hardwired to connect with others, it's what gives purpose and meaning to our lives, and without it there is suffering."
>
> — Dr. Brené Brown, *Daring Greatly*

Others are Internet superstars.

I Am Kind Of A Big Deal On The Internet

If we are a big deal on the Internet, are we really that big of a deal? It depends on our definition of success. If we define success as fame — that is, arm's length known-ness — then yes. But not if we define it in terms of relationships, the quality of people we love and care for and who love us back.

How we define success matters. Advertisements are constantly telling us: "You are not thin enough, young enough, rich enough, and you're not sexy enough. If these are your values, it's probably because you have been watching too much television," says media expert Dr. Read Schuchardt, in his series "Living in a World with No Off Switch."

His advice? Burn your television.

"And voilà, problem solved. You will now have an extra 35 hours a week in which to lose weight — you're thin enough, workout — you're young enough! take a second job — now you're rich enough, and meet your wife — now you're sexy enough! In other words, if you eliminate the voices that are telling you you are not these things, you will actually become those things. But guess what? You'll also recognize: 'Wait. Those are crappy values.' You will actually then be free and mentally liberated, emotionally liberated, psychologically liberated, to ignore the mass culture's spiritually disillusional values and . . . actually get a life."

Consider this: every country has its own celebrities. You don't know the hottest actor in the Ukraine, the biggest reality star in New Zealand, the top investment banker in Hong Kong, or the preeminent ballerina of Colombia. But there is someone at the top of every industry in every nation of the world, and it is changing every second, and we are chasing *that.*

Chapter Challenge:

How do I define success? (Limit your answer to three things.)

____

____

____

Stressified

Too rarely do we hear about the physical cost of high-stress, high-powered competition, and the ensuing loss of the inner life. Unattainable employer expectations combined with no clear tech boundaries, have created a whole new level of workaholism where exhaustion and productivity are our measure of self-worth.

Alexa, a senior software manager at one of the world's top mobile phone manufacturers, says she feels constant pressure to work beyond her contract. "When I joined here, our slogan was *'Always on.'* My manager always responded to emails outside of business hours, stayed late, etc. Informally, she set a bar to do the same. There is a lot of unwritten culture here to be responsive to emails at all times of day. If you want to get in on the discussion, you better answer as quickly as possible — whether you're in a meeting or about to go to bed."

> "The world [we have created] is based on the notion that we are all separate. Like our education system valuing independence and competition, business based on scarcity and competition, and our towns and cities based on the idea of separation. So we fashion our world so that we are significant at someone else's expense."
>
> — *I Am* documentary

She joins the growing majority of people who say that the idea of being "off-the-clock" has disappeared. "We're constantly connected, and it is hard just to shut it off and focus on the reality in front of you," says Alexa. "I mean, I don't just watch TV anymore. I watch TV while reading my newsfeeds/social feeds.

"I am hearing more and more about the effects of being always on, the impacts of screen time, and the importance of downtime. I recently heard of some laws to make it illegal to email outside of business hours. Something like that type of legislation seems extreme to me, but it will take a lot of individual effort to make change. As a manager, I try to set an example of what I expect outside of office hours. I try to be open with people about what to expect. I try not to live in my inbox while at work. In the car, I keep my device in the back seat. At home, I keep it in my purse (at least until the kids have gone to bed!).

"The less you respond outside of business hours, the less you will be called on to do so. But will that impact advancement?"

How do we claim our boundaries and redevelop our focus?

Often our "get ahead" career culture is at odds with our deeper values such as caring, compassion and kindness. For many, this can lead to a deep loneliness and sense of alienation in the workplace. Having little free time or attention, we develop tunnel vision; important concerns take a back seat to the demands of simply surviving in the career culture.

Is it worth it?

Shifting to a Culture of Support

Lisa Ferguson, a successful interior designer, was an early Twitter adopter. Her business grew steadily along with her online presence but, sadly, she's also become a target for what she calls, "mean girl behavior."

"Even for those mostly secure in who they are, comparing your reality with another's 'highlight reel' on social media can draw out insecurities and jealousies," says Ferguson. "So, many show up in social media with 'lack' mindsets and deep-rooted insecurities. They appear to play nice and be supportive, but when egos and jealousy rear their ugly heads, almost in no time they convert to bullying others. Some would pull the same aggressive behavior offline, but being behind the screen has fostered another type of bully who feels bolder when their target isn't live in person.

"Often, with an ultimate agenda of feeding ego, insecurities or creating celebrity status for oneself, this means consciously or unconsciously tearing others down to build oneself up. My energy shifts just by reading

> "Rather than the attitude of 'how did you get that client?' why not have an attitude of abundance, gratitude and encouragement?"
>
> — Lisa Ferguson, founder of Strengths Mentor

> "As soon as we start selecting and judging people instead of welcoming them as they are . . . we are reducing life, not fostering it. When we reveal to people our belief in them, their hidden beauty rises to the surface where it may more clearly be seen by all."
>
> — Jean Vanier

caustic comments, so I am intentional to remove people who tear others down," says Toronto-based Ferguson.

And she's completely changed her professional life in the process. Today, Ferguson runs a new business called Strengths Mentor, a service that helps people take a Strengths Finder assessment and map out meaningful personal and professional livelihoods.

"I love to elevate, support and cheer on peers when they do respected work or achieve victories. I believe championing others is the right thing to do. I often find people want to hire me because they respect and admire that other-centered quality. Go figure that supporting others may be a key secret to success! Many people show up with a 'what's in it for me' mindset? Fair enough. I prefer to show up with a 'mutual win' perspective."

It's Easy Being Mean

> "Our relationships are being destroyed by a bad way of life. We use speech so abstract, so far removed from anybody's experience, that it is virtually out of control; *anything* can be said if the speaker has foolishness or the audacity to say it."
>
> — Wendell Berry (*What Are People for?* "A Remarkable Man," p 20)

The things people write in the anonymity of their basements can embarrass even the crassest among us. We can mutter to ourselves online and people flood in with approval. It's as if the mediated message gives us permission to speak the unspeakable.

There are a lot of celebrities on Twitter, and some people feel very comfortable using that platform to insult those celebrities. What we don't see is the pain those words can cause. In an effort to get people to pause and reconsider before posting something awful, late night TV host Jimmy Kimmel asks famous people to read mean tweets aloud on his show, reminding us that words have the power to hurt.

Bullying online is a problem within the tech industry itself. Online gathering spots like Reddit, Hacker News and 4chan, where people often post anonymously, can feel like hostile territory, reports the *New York Times,* especially for women, says Lauren Weinstein, a consultant for Google working on issues of identity and anonymity online. "Many women have come to me and said they basically have had to hide on the Net now. They use male names; they don't put their real photos up, because they are immediately targeted and harassed."

A colleague of mine, Nadine Silverthorne, online editor of *Today's Parent,* publicly shared a hesitation she had about vaccinating her kids. Within minutes of the story being posted on the *Toronto Star* newspaper's website, people were up in arms calling her every despicable name in the book. She came over to Facebook to write the following:

"Please think before you post your angry tweets. I get that we're all afraid that our kids will get measles or the whooping cough, but when we come out as a collective 'us' with flaming torches and yelly voices because of fear, we feed a beast that can lead us to terrible actions. If you want to persuade people to come over to your side, *this is not the way.*" (Emphasis mine.)

Surely, it is not.

> Social media platforms, where people meticulously build their public appearance, often become bullhorns for offensive comments.

> "It has been said that astronomy is a humbling and character-building experience. There is perhaps no greater demonstration of the folly of human conceits than the distant image of our tiny world. To me, this underscores our responsibility to deal more kindly with each other."
>
> — Carl Sagan

Mindfulness Leads Us Away from Judgment

In her work studying the illusion of control and decision-making, Ellen Langer, professor of psychology at Harvard University and a Guggenheim Fellowship recipient, has discovered that mindfulness makes us less judgmental about others. "We all have a tendency to mindlessly pigeonhole people: He's rigid. She's impulsive. But when you freeze someone in that way, you don't get the chance to enjoy a relationship with them or use their talents. Mindfulness helps you appreciate why people behave the way they do. It makes sense to them at that time, or else they wouldn't do it."

We all have our quirks and default ways of being. Langer explains that if we were to rate our own character traits — the things we most valued about ourselves and those we'd like to change — we'd discover a great irony. "The traits that people valued tended to be positive versions of the ones they wanted to change." For example, Langer values being spontaneous but also sees herself as being too impulsive. "That means if you want to change my behavior, you'll have to persuade me to not like spontaneity. But chances are that when you see me from this proper perspective — spontaneous rather than impulsive — you won't want to change me."

The Delight of Self-Forgetfulness

One morning not long ago, I found myself bouncing between websites — my blog, Facebook and Twitter accounts — with the sole intention of managing my online persona. An add here, a tweak there. That week, I'd been assigned six new stories, invited to participate in an anthology and was busy juggling the contract for a September job offer. By 11 AM I was exhausted with myself. The Internet, for easily self-obsessed "busy little doers" (as Kathleen Norris calls them) like me, is an online prison of our own making.

In his book *The Narnian: The Life and Imagination of C.S. Lewis,* Alan Jacobs tells us that in most children but relatively few adults, at least in our time, is the willingness to be delighted to the point of self-abandonment. "This free and full gift of oneself to a story [when reading] is what produces the state of enchantment. But why do we lose the desire — or if not the desire, the ability — to give ourselves in this way?" he asks.

"Those who will never be fooled can never be delighted, because without self-forgetfulness there can be no delight."

Inspired by these words, I decided to try spend a month attempting to forget about myself, once again leaving behind my online tinkering to immerse myself in the story unfolding all around me.

As I write this, I am on a small hobby farm in British Columbia's interior, sitting cross-legged with a laptop propped atop an old knit blanket. Nearby, my cousin Liz's brood of four are zigzagging the property with mine, each clamoring about on the straw and sand.

Where I'm sitting is a bit messy. I'm leaning against an old wood fence where a cobweb is stretched inches from my elbow. The morning sun is dancing through the expertly crafted web as it gently pulls and pushes with the wind. Through it I spy a small gray field mouse. In and out he comes from his tiny burrow, stopping to look, acknowledging my curious eyes. I can't remember a time when I've had both the good fortune and the time to stay attentive to a wild rodent at work.

My city self is quick to brush away the cobwebs; usually, I live in a steady state of panic. It is my chief sin, this busyness, and friends and family know it full well.

But here, where I have stopped to take pause this morning, my eyes are open. When I stepped off the plane two days before my little boy's first birthday, I made a promise to myself: this month would be different.

My cousin Liz made it easy. "I can't wait for your visit," she wrote the week before our departure. "No cement-clad high rises or spider-legged byways here. Just pastures and unicorns and free-range children." It's not every day you receive an invitation to frolic in fields among the anthropomorphic creatures of your childhood imagination. We've come to my cousin's home for a family reunion. The unicorn is really a pony, but my kids don't need to know.

Yesterday I spent the better part of an afternoon in the local Auxiliary Thrift Store — a busy volunteer-run shop filled with small-town minutiae, gently-worn blouses and a handsome collection of used books — where I filled a bag for $5 and change.

As I stepped out into the August sunshine and wandered down Main Street, I found myself standing face-to-face with my old roommate Sarah Waters — the same Sarah I befriended at a political convention and who emceed my wedding reception. Here we were, hundreds of miles from our respective homes, shrieking with laughter in the center of this old retirement town.

I found myself wondering, if we'd both checked in on Foursquare and found each other that way, would it have felt the same?

> "To let ourselves sink into the joyful moments of our lives even though we know that they are fleeting, even though the world tells us not to be too happy lest we invite disaster — that's an intense form of vulnerability."
>
> — Brené Brown

Here, in the arms of my friend, here with my palms in the raspberries and feet in the sand, I am rediscovering the delightful foolishness of forgetting myself.

The Grass Is Always Greener through a Filter

Are your online engagements sapping your energy? Do you feel starved for time, distracted, ill at ease? If so, step off. You began using these tools to help you, to bring enjoyment, access information, feel connected. If they're enjoining you to compete, dragging you down, then ditch them — at least temporarily.

Chapter Challenge:

Take a moment to think about a person, colleague, old classmate, celebrity or random stranger you find yourself comparing to online. Write their name here:

____

Now think about the person you know, in real life, that is the closest thing to them. What are their flaws, their limitations? Write them down here:

____

Let me break this down: Every person is human. Everyone has limitations. Beyonce's baby pukes on her, and Justin Timberlake and Jessica Biel fight, I promise.

Forget about them and get on with it. Life is waiting outside the door.

Chapter Challenge:

I will forget about what other people are doing and just do my thing.

Signed,______________________________________

9

Coming Close

Trust

It was a cold Toronto morning when I pulled on my salt-stained boots, crunched over to our army-green hatchback, and slowly drove through a clamor of white to meet up with my theatre-actress aunt on the east side of town.

As I swung the cafe door open, two small children scurried about my knees, and I immediately did a double-take on a woman with pulled-back dirty blonde hair. Though I misremembered her name, it was in fact who I thought it was: Eva, from my small Vancouver middle school. She had been a grade younger than me, the daughter of the school music teacher, known for her sweet personality and tumble of wild blonde curls. Here she stood before me 2,718 miles from home. I learned she'd been living in the city for five years, now has a couple of kids and teaches music part-time while her husband works for Luminato, the annual avant-garde arts and creativity festival. We made plans to get together at the same eatery where she leads a musical circle time for children.

The wonder of this encounter got me thinking: when we abandon the Internet, even for a short time, we abandon ourselves to trust, to the belief that moments like this will happen. It was magical. Neither of us had a hand in it, but *Providence* did.

Online there is always someone initiating, but here in the "real world" we are left to these chance, or not-so-chance, encounters. It's a wonderful feeling.

The Necessity of Needing

There was a time when people relied upon each other's skills rather than technology for everything. Babies were birthed by skilled midwives, and barns were raised by entire communities. Word of these events spread by word of mouth, person to person, at community events such as church and town meetings, or when passing one another along the road.

I experienced this for a short time on a small island in the Pacific Northwest. My husband, one-year-old daughter and I moved to a vacant house on the ocean. The home, complete with an enormous north-facing wall of glass, was situated on a parcel of family land belonging to our friends, the Cowpers, who lived around the bend in the original family cabin. They were the ones who taught us how to close the makeshift latch to keep the bears (who had swum to the island) out of the garbage. Family members checked in with us daily. *Did we have enough fire wood? Did we find the Septonic for the toilets? Had we squashed all the furry giant spiders? Had we heard about the Knick Knack Nook — the new thrift shop — on the isle?* Sure, we had a telephone, but taking a pass by the house seemed to make more sense. We were neighbors in a remote place. The winter winds were coming, and well water pipes had a tendency to freeze. We were going to need each other.

Our culture values autonomy, independence and success above all, but therapists, educators, parents and spiritual directors agree that trust, rooted in intimacy and need, is integral to human relationship. But trust is hard to find where we spend the lion's share of our time: on the Internet. A Google search returns more than one million hits in 0.19 seconds but there's nothing there about interpersonal trust. What you'll find instead is *Trust* the band, *Trust* the movie, and *American Trust,* the bank.

> The word *trust* originates from *traust* meaning "help or confidence" from Middle English, probably of Scandinavian origin. It is akin to Old Norse traust and Old English trēowe, meaning "faithful."

"[Trust] increases security in a relationship, reduces inhibitions and defensiveness, and frees people to share feelings and dreams," report researchers Stinnett and Walters. Unfortunately for us, without careful discipline, trust is relegated to the sidelines as technology speeds and our sense of control increases.

So, where do we go to cultivate these trust-based relationships?

When I stepped offline for a month, I had no choice but to trust in chance encounters. And when I ran into a girl from my middle school two thousand miles away one day in a snow storm, it was magical. And when I accidentally locked my sleeping newborn baby inside my house and couldn't Google search a locksmith, I had to run to my neighbor's house where she gave me what I needed above all: comfort.

Craigslist Joe

Screens allow us to live in silos. Many of us are too busy fondling our handhelds to engage with the person next to us, desensitizing us to the needs of others and reducing opportunities for trust and wonder. When was the last time you struck up a conversation with a stranger on the subway platform or reached out to a neighbor on the street?

Chance encounters seem to be a thing of the past because online there is always someone initiating. Technology give us a sense of control, and where control reigns, whimsy flees. Yet the hunger for intimacy remains. So much so that today you can "feel" your partner thousands of miles away using a combination of specialized underwear and a downloaded app. And the teenage girls next door tell me that even though they communicate almost exclusively by text message, they still expect to be asked out in person.

> Thanks to our smartphones, it's easier to turn to Safari than our neighbor for help.

One man showed us that it is possible to rely on the kindnesses of others in the 21st century, and that this reliance has profound implications: intimacy, expressed need and trust are inextricably bound.

> "In a time when America's economy and sense of community were crumbling, one guy left everything behind — to see if he could survive solely on the support of the 21st century's new town square: Craigslist.
>
> As of recent, the United States found itself in one of the most precarious financial meltdowns in modern history. News programs spoke of the worst economy since the

> Great Depression and demise of the American Dream. Unemployment was soaring and millions were losing their homes. Rather than banding together and helping one another, people started pointing fingers and casting blame. Many feared the sense of community that had once carried us through tough times had dissolved into an attitude of 'every person for himself.' Many were skeptical that today's self-involved society would be able to weather the storm without its traditional social supports.
>
> It was in this climate that 29-year-old Joseph Garner cut himself off from everyone he knew and everything he owned, to embark on a bold adventure. Armed with only a laptop, cell phone, toothbrush, and the clothes on his back — alongside the hope that community was not gone but just had shifted — Joe lived for a month looking for alms in America's new town square: Craigslist. For 31 December days and nights, everything in his life would come from the Craigslist website. From transportation to food, from shelter to companionship, Joe would depend on the generosity of people who had never seen him and whose sole connection to him was a giant virtual swap meet.
>
> Would America help Joe? Could he survive with nothing, apart from the goodwill of others?"
>
> (Source: www.craigslistjoe.com)

He found, indeed, he could.

Online, need is usually relegated to the sidelines. But Joe's needs created opportunities for creating relationships, fostering trust and cultivating intimacy between the unlikeliest of characters. Craigslist, especially in the case of Joe, is a fantastic example of how the Internet can help us communicate need and share resources.

The key is intentionality.

Intimacy and Trust

"Intimacy relies on safety, patience, mutuality, respect, constancy, and no secrets," writes Randi Kreger in *Psychology Today*. "Without healthy self-disclosure at the right time, there can be no intimacy. And that takes honesty about who we are and how we feel. The more intimate you are, the safer you feel and the more worthwhile the relationship."

Real intimacy is cultivated through vulnerability, moored in trust and sustained by understanding and feeling understood. "Vulnerability is the core, the heart of meaningful human experiences," says Dr. Brené Brown, research professor at the University of Houston Graduate College of Social Work. (Source: *Daring Greatly,* p 12) To be vulnerable is to dare greatly, as we intentionally express our needs, shame and shortcomings, acting in hope that we will be accepted as we are.

It is impossible to be vulnerable, and therefore intimate, without expressed need. Intentional vulnerability, as in the case of Craigslist Joe, creates an opportunity for others to draw closer, to rise to the occasion to help.

According to psychiatrist James Masterson, author of *Search for the Real Self: Unmasking the Personality Disorders of Our Age,* sharing what is deepest and most real about ourselves followed by mutual sharing is of vital importance to a sustained, mutually satisfying relationship. Unfortunately, he says, this is something narcissists — those who have an excessive interest in oneself, with a grandiose view of their own talents — have a difficult time doing.

Facebook alone is a billion narcissists strong (including me). The root cause of this narcissism epidemic is what Dr. Brené Brown calls "the shame-based fear of being ordinary."

"I see the cultural messaging everywhere that says that an ordinary life is a meaningless life . . . And I see how kids that grow up on a steady diet of reality television, celebrity culture, and unsupervised social media can absorb this messaging and develop a completely skewed sense of the world. *I am only as good as the number of 'likes' I get on Facebook or Instagram."* (*Daring Greatly,* page 23)

Trust is experienced in close proximity. It is within our ordinary lives, within the intimacy of our nearest relationships, that trust and intimacy can be forged.

But the trust must be real.

One of the buzzwords in the world of social media is "trust." We are told that in order to foster personal or professional online relationships we must gain the trust of others, demonstrating through our content and candor that we are worthy of this trust. But at the core, even for the best-intentioned bloggers, motivations are hidden. We are building a brand. We are selling a product. We are presenting a particular persona.

In his 2010 book, *The Authenticity Hoax,* Canadian journalist Andrew Potter explains how since the entrenchment of technologies such as the Internet, the exclusivity-based cool brandished by 20th-century cultural elites gave way to what he calls "conspicuous authenticity." Our pursuit of authenticity is in many ways the pursuit of a mirage, and Potter argues that the pursuit of it is ultimately not just futile, but destructive.

According to prominent sociologist Niklas Luhmann, one of the primary challenges of our day is to examine closely how the rapid progress of technology — particularly information technology — has impacted our understanding and experiences of trust.

> We have replaced conversations with connections.

Genuine intimacy in human relationships requires dialogue, transparency, vulnerability and reciprocity. The verb "intimate" means "to state or make known." The activity of intimating (making known) underpins the meanings of "intimate" when used as a noun and adjective.

Is the very design of communication technology and the mediums therein a hindrance to the intimacy necessary for relationship? How can we work within them to foster the intimacy we need?

Social Etiquette 101

Have we lost all social etiquette? A page in an etiquette guide from 1888 reads:

- You should dress quietly at your own entertainments, thus avoiding the possibility of eclipsing your guests.
- You should speak without arrogance to those serving you. They may be as well born as yourself.
- You should strive to live down all false and evil reports, but never to contradict them. ☞

- If you have been sick yourself, say as little about it as possible, and never allude to it at the table, where you will receive little sympathy and perhaps render yourself offensive to all who hear you.
- You should shun boasting. It is vulgar.
- You should reject dresses that are too low necked and those that have almost every vestige of sleeve cut away.
- You should, if old, show consideration for the faults and follies of the young. If young yourself, be gentle and respectful to those of riper years.
- You should never take dogs with you, nor pets, unless they are specially invited.
- You should never bang doors, especially entrance doors.
- To laugh heartily, or to whisper unfavorable remarks during the performance of a concert or a play is a rudeness of which no gentlewoman would be guilty.

And, here's the clincher: *Be careful not to show your ankles, however pretty they are.*

Face-to-face, we understand the etiquette. The more direct our communication, the more intimate and meaningful it feels. This is why lovemaking and our most intimate conversations happen one-on-one, without distraction or interruption. In love, we offer our full attention.

Spreading the News

Before I stepped offline, I had grown weary of Facebook mediating my relationships. I had missed a friend visiting Toronto because she messaged me twice on my Facebook "wall," and I didn't check it in time. I felt sad and frustrated with her for not communicating with me more directly by phone or email. I'd also found out about the birth of one my best friend's children by way of social media. The lack of intimacy in this sharing was cause for grief. And I know I've committed the same *faux pas*. I let word get out about my nephew's birth while my parents were away on holiday. Rather than finding out via a call or email, my stepmom stumbled across the news on her Facebook home page. Ouch. They were none too happy, and it was with sheepish trepidation that I delivered my apology.

Practice what you preach, Christina. Practice what you preach.

We have no delay buttons. We throw things out online for others to find, putting the onus on another person to know what's going on in our lives. It can feel passive aggressive to communicate this way. There is also no way to know who has read what. Did my mom read my last blog post? Did Sarah read my last few status updates. Is sharing these things again in person redundant, boring, helpful, annoying?

While offline for 31 days I had to initiate or hope for direct contact from others. Before that, it was not uncommon for me to check in with people by firing an email, text or Facebook message that read: "Hi. Thinking of you . . ." then relying on their recent posts to paint a story of their lives.

But do our online personas paint the real picture? Most of us do not pour out our needs and challenges, big or small, on Twitter and the like. Those who do are often met with scoffs or offline murmurs about oversharing. A person can't win.

One week of my Internet fast brought two long phone conversations with my best friends on the West Coast. Through them I learned so much more than the sum of their online updates. (Full disclosure: they initiated both calls.) *Thank you, friends.*

> The more direct the communication, the more intimate it feels.

In many ways, the Internet has served as a blessing for my very large immediate family. I am one of eight kids now spread out as far as Las Vegas and small town Australia. A single email can connect us all to the same photos and stories. The more direct the communication — an email vs a Facebook post, for example — the more intimate and meaningful it feels.

This is why lovemaking happens one-on-one.

Intimacy and the Internet

"North Americans are bringing their smartphones into the bedroom in droves. Yes, texting while having sex," reports *Psychology Today*. "A recently released study indicated one in ten participants admitted to having used their phone during sex. As far as young adults, ages 18 to 34, make that one in five — 20 percent." (That would be like the 1980s equivalent of writing a post-it note during sex. Kind of impressive, when you think about it.)

Sex educator Kim Sedgwick, co-founder of the Red Tent Sisters, is a strong proponent of leaving phones out of the bedroom.

"The question about screen time impacting couple's intimacy is something that comes up in my practice all of the time. In my experience talking to clients, screen time has the potential to cause friction in romantic relationships. Many employers expect you to be available by email even during off hours, which means it's hard to get out of 'work mode' even at home. For others, the issue is social media and texting — conversations are expected to happen in 'real time' which makes it easy to feel like you *have* to reply immediately. Whether a partner is actively online or simply distracted as they glance at their phone every five minutes, the result is often the same — a sense of disconnection and devaluing of real intimacy."

For Sedgwick, this comes down to setting clear expectations and trusting one another to follow through:

> "That said, I have found that some couples have managed to use screen time to their advantage. For instance, my sister Amy is currently in a long-distance relationship with her husband, and she's found Skype to be hugely helpful for maintaining intimacy. Sometimes she and her husband will have a Skype dinner date or they'll listen to a song together. These activities provide a sense of prioritizing time with one another, even if they can't be physically connected. Similarly, I know lots of couples who use text as a way to build anticipation about spending time together. So, I don't think screen time is inherently detrimental to intimacy. The question is how (and when) it is used, which requires open communication and setting boundaries.
>
> When it comes to phones in the bedroom, not only do they negatively impact intimacy by providing a distraction, but there is all kinds of evidence about how light from electronic devices disrupts sleep (plus the stress that can be caused by checking email before going to bed). If you're overtired and stressed, you likely won't be a very enjoyable

companion. So, for lots of reasons, I suggest checking your phone at the door.

I think any time you make a conscious effort to be with your partner without being tethered to technology — whether for an hour, or a day — makes a huge difference. You're sending the message that you value time together; they're more important than how many likes you got on your Facebook status. In an age where we're constantly multi-tasking, it feels pretty special to be able to trust your partner to focus entirely on you."

> "Trust is a beautiful form of love. When we are generous, we give money, time, knowledge. In trust, we give ourselves." — Jean Vanier
>
> — *Becoming Human*, p 28

Trust is essential to true intimacy and a product of vulnerability that grows over time. It requires attention, work and full engagement. It asks all of us.

What Do We Lose by Waiting?

Here in the 21st century, we bow to the god of instant gratification. It is called out of us on Instagram, Twitter and Facebook: *What's on your mind?* It's delivered to our inboxes, placed in our newsfeeds, enabled by Amazon and propagated by every preferential banner ad.

> "Easy is better. But as machines do more work for us, we do less; we're less capable on our own. More is better. But as machines store and organize more, we get sloppy, forget our friend's phone number, birthday, heartfelt concerns."
>
> — *Geez* magazine, issue 20

The longer I navigate the demands of the Internet, the more grateful I am for my children. They save me every day. At each juncture, their very tangible needs crash against my frailty, and I must reach out to meet them. Without the demands of these little people I would easily slip into spending days the way I spend my nights: glued to the screen. Netflix is my gateway to relaxation, Facebook my voyeuristic portal of delight. Left to my own devices, I'd drain the currency of my life down Alice's rabbit hole.

Instead, I am forced into the present, my children draw me close.

Intimacy: A Slow Build

Sometimes we don't choose to go slow.

We're sideswiped by sickness, bedridden, lost to the sheets. Some of us have bowed to the quotidian nature of home economics: bending low to clean floors and collect toys, over and over and over again.

Myself, I feel the grip of *slow* most firmly on days spent inside: long hours passed with small children on the sixth day of rain.

Although it can sometimes feel like I'm standing on the brink of madness, it's in this slowness, the repetition, writer Kathleen Norris tells us, that we can begin to recognize and savor the holy in the mundane circumstances of daily life.

In her slim volume, *The Quotidian Mysteries,* the author writes: "The fact that none of us can rise so far in status as to remove ourselves from the daily, bodily nature of life on this earth is not usually considered a cause for celebration, but rather the opposite . . . We want life to have meaning, we want fulfillment, healing and even ecstasy, but the human paradox is that we find these things by starting where we are, not where we wish we were. We must look for blessings to come from unlikely, everyday places." Like laundry.

The Internet and other technologies allow us to escape the inner gaze while painfully slow moments — peeling carrots, tending to a sick child, waiting for a bus — necessitate reflection.

"Workaholism is the opposite of humility, and to an unhumble literary workaholic such as myself, morning devotions can feel useless, not nearly as important as getting about my business early in the day. I know from bitter experience that when I allow busy little doings to fill the precious time of early morning, when contemplation might flourish, I open the doors to the demon of acedia. Noon becomes a blur — no time, no time — the wolfing down of a sandwich as I listen to the morning's phone messages and plan the afternoon's errands. When evening comes, I am so

> A life of trust, a life free of false sense of control, is freedom. Perhaps that is the key: viewing the hours of the day with honor, seeing them as precious and treating them as such. Each moment is given as a gift; the terminally ill know it full well, yet we healthy, upwardly-mobile types drain the day of its possibility with our "busy little doings."

exhausted . . . It is as if I have taken the world's weight on my shoulders and am too greedy, too foolish, to surrender it to God," writes Norris.

> We need moments free for contemplation. Few of us, for example, can pray or compose a poem while watching television or reading Reddit.

As our technology has sped, so have our heart rates. Our thoughts, once paced by stories spoken, the rhythm of our steps, the text on a page, are rushed by an onslaught of information with which none can keep pace. In the absence of technology, in the presence of nature and people, we find a different kind of story.

In Wes Anderson's 2012 film *Moonrise Kingdom,* two young lovers begin their journey in a wild summer field. The orphaned boy and his troubled friend have kindled their friendship by way of letters, and the connection deepens as Sam teaches Suzy to pitch camp and she reads to him by moonlight. The two plan to flee their small New England town which causes a local search party to fan out and find them.

My aunt Astrid recommended *Moonrise Kingdom,* promising me I'd love it. She was right. I don't keep up with Pinterest, and I have trouble staying on top of Facebook and Twitter, so chances are good I would have missed the movie had she not made a point of telling me about it.

In *Moonrise Kingdom,* Sam shows Suzy how to catch a fish and fry it. Suzy shows Sam the magic found in her suitcase full of books. These lessons stay with us, the ones that unfold in relationship, the ones that then lead to repetition and repetition to mastery.

The other day I stood barefoot in my friend's kitchen. While describing the nonfiction she's crafting, her hands expertly cradled a red pepper. I watched, enraptured, as she carefully sliced around the top, leaving the seeds intact, and carefully julienned the thick juicy skin. I'd always known I wasn't doing it quite right. And here, without words, Jen showed me the trick.

My grandmother showed me the secret to the best French toast. Every morning, after my single mom dropped us off before school, she'd butter a stack of bread, soak each slice in egg and cinnamon, then place each piece in a pre-buttered pan. When I was ten, I learned that double-buttered French toast makes the world better, and I continue the tradition with my family.

Educators understand that learning takes time. Yet somehow, when we graduate from the classroom, we forget this. The Internet has made it possible for us to Google search our questions at a moment's notice and, while Wikipedia and Ask.com are brimming with answers, I think we often forget that the lion's share of learning doesn't travel well along the information highway — things like building tree forts with siblings, swapping marinade recipes around the barbecue, learning a new sport with a friend.

New Yorker Tim Kreider says he just wants to spend time with his friends but, he writes, "almost everyone I know is busy. They feel anxious and guilty when they aren't either working or doing something to promote their work . . . I recently wrote a friend to ask if he wanted to do something this week . . . But his busyness was like some vast churning noise through which he was shouting out at me, and I gave up trying to shout back over it."

"I was a member of the latchkey generation and had three hours of totally unstructured, largely unsupervised time every afternoon, time I used to do everything from surfing the *World Book Encyclopedia* to making animated films to getting together with friends in the woods to chuck dirt clods directly into one another's eyes," recalls Kreider in his *New York Times* op-ed, "The 'Busy' Trap," "all of which provided me with important skills and insights that remain valuable to this day."

The best learning happens in the wide open spaces: physically, spiritually, mentally and relationally; disciplining ourselves to carve out these spaces may be the most valuable lesson we can learn.

"I did make a conscious decision, a long time ago, to choose time over money, since I've always understood that the best investment of my limited time on earth was to spend it with people I love," says Kreider. "I suppose it's possible I'll lie on my deathbed regretting that I didn't work harder and say everything I had to say, but I think what I'll really wish is that I could have one more beer with Chris, another long talk with Megan, one last good hard laugh with Boyd."

Like Kreider, I'd rather learn my lessons firsthand, sitting in my backyard taking notice of the trees and listening to my friend Avital.

Turns out this intimacy is also essential to our mental health. Go figure.

Chapter Challenge:

The next time you have an urge to post something online, share it instead with a friend or family member directly. Call and tell them your story. Take over some photos they can't find online. Or, if you are even more ambitious, write a letter. Tell them about life. Draw them a picture. Tuck in a quote. Get creative. Show the love. Notice their response. Pay attention to how you feel.

10

Introduction to Part Three

The Way Forward

Media theorist Marshall McLuhan coined the phrase "the medium is the message." This idea has never been more obvious than in the age of the smartphone. Our digital devices dominate our every day — affecting our work, impacting our intimacy and shifting our thinking. For all the affordances of our new communication technologies, we haven't established healthy norms or habits as a culture, and it's costing us our time, creativity, energy and relationships.

> A life of trust, a life free of a false sense of control, is freedom.

Ninety percent of adults and teens have cell phones, and the smartphone number climbs every day. The number of mobile users in Africa surpasses that in both the US and Europe — and in some African countries, more people have access to a mobile phone than to clean water or electricity. Our Internet use has shifted profoundly from stationary computers to mobile devices that travel with us everywhere. We are always on, never off.

A neighbor of mine recently shared with me that almost every time she sits down to feed her baby, her first instinct is to grab her phone to check messages or catch up on Facebook. It grieved her when she thought about how many times she'd looked at her phone rather than into her child's face.

During my Internet fast, I made calling my grandma, who lives on the other side of the country, a priority. Instead of checking my email for the 16th time, I would pick up my phone and call her. One half-hour conversation turned out to be the last lucid one we would have. She passed away later that year.

I am no great example. When it comes to wasting time on an iPhone, I am the worst offender.

Here's how we can begin to make people our priority:

First, let's silence and put away our handheld devices and look at people in the eyes. The person in front of us is the most important part of that moment. Let's make them feel that way, and hopefully they'll do the same for us.

Together, let's make social smartphone fondling a faux pas.

Second, let's reclaim sacred spaces. When getting home from work, let's begin dropping our phones at the door with our keys. If you have a family, you will be more present to them, and they will love you for it. And if you are single, you'll carve out a quiet haven at the end of your day, and you will love you for it.

Let's make mealtime about conversation and being in the present moment.

Third, honor the holy hours. We have a window of time when we first wake up that will set the course for our entire day. Spend the time reaching for a loved one, reading a scripture, meditating on a goal, or simply sitting in quiet. The time is yours, and it is currency.

Since the late 1990s, we've quickly embraced email, cell phones and search engines with wild abandon, but our habits and disciplines have not caught up. Disconnecting gives us the space to create, allowing us the mental real estate necessary for invention.

> On the horizon is a technology that could eclipse the iPhone. It's a slim wearable computer called Google Glass. When wearing Glass, how will we be able to tell if someone is truly looking at us instead of shooting a video or checking their stock portfolio? We won't. Not unless they set them aside and give us their full attention.

We don't use a screwdriver to butter toast. We don't use a knife to write a love letter. Use the Internet like any tool. Take it out for a specific purpose and then put it away.

We are living in a constant state of information overload and a vacuum of joy. We have too much information and not enough wonder. We need a new equilibrium, a mindful approach to our use of technology, and we have the opportunity to choose it now.

Three Keys

I believe there are three keys to cultivating a meaningful existence in the 21st century; they are: Embrace Weakness, Practice Renunciation and Be Known.

First, in admitting weakness, we confess our need and come closer to one another.

Second, is renunciation. In letting our yeses be yes and our noes be no, we form commitments, and in these limits we find our meaning and joy.

Third, we must be known. We matter most to the people in our lives who truly know us, and, in honor, we must devote our lives to them. We do not discover who we are in a solitary state; we find it in mutual dependency, in learning through belonging.

C.S. Lewis said "Practice the discipline of planned neglect. Discipline yourself to pursue the better." We live in a world full of computers and people. Let us together pursue the deeper, harder work of putting people before our technologies. Our lives will be the better for it.

11

Reorienting a Life

Learning Our Lessons Longhand

The difference between a life lived actively, and a life of passive drifting and dispersal of energies, is an immense difference. Once we begin to feel committed to our lives, responsible to ourselves, we can never again be satisfied with the old, passive way.

— Adrienne Rich

Watchful Reverence

It's May. Six other women and I have traveled to Prince Edward Island to a turn-of-the-century hotel to spend three days exploring writing and inspiration with Sabrina Ward Harrison, beloved artist and author of the cult favorites *The True and the Questions* and *Spilling Open,* both excerpts of her personal journals. The Highlands is an old waterside hotel whose rooms have housed royalty and the likes of Reverend Billy Graham. You can find their names penciled in the guest book downstairs, next to a collection of seashells and stacks of well-thumbed *LIFE* magazines.

When Sabrina breezes into the sitting room, she greets each person with a sincerity that burns, a kind of presentness, a "watchful reverence for the moment," which she references over and over during our stay.

Later, receiving a compliment on her sunny complexion, she shares her secret: "It's self-tanner." And because it's a brand you can only get in Canada, she's stocked up for her trip home to Silverlake, California. She could have easily slipped by the comment but instead she spilled the truth: she burns terribly and the tanner helps. It may sounds like a trivial subject, but for Sabrina, these moments of confession are her life's work.

Old magazines are strewn throughout the Highlands's three floors, and I can't help combing their pages. A 1955 *National Geographic* depicts the allure of the Gulf Islands off the west coast of Canada: "The calm leisurely atmosphere of the islands works its charm upon people from all walks of life. The person in worn clothing whom the newcomer mistakes for a beachcomber or country bumpkin as often as not proves to be a famous scientist, novelist, admiral, politician, or other notable . . . The sympathetic outsider may find himself being initiated into what may be one of the last truly civilized ways of life remaining in English-speaking North America. Once that happens, it is unlikely that he will ever want to leave."

Prince Edward Island's appeal resides in its people and natural beauty: the steady calm of the island air whistling through the birch trees; the burst of plover, finches and jays that begin their daylight calling at 4:30 AM; the rust red roads with their truck tracks drawing a path to our doorstep.

Becoming Yourself

I am sitting on the front stairs of the main house. In front, four crooked trees congregate like an outer hearth. The twisting white, worn branches are the sort you'd find in the Haunted Wood of Anne of Green Gables's imagination. The beaked chirps, caws and whistles blend into a symphony of spring. Behind me, seven women chatter at the breakfast table, while passing around fresh-baked muffins, boiled eggs and preserves.

In the adjacent old dance hall, Sabrina, dressed in a vintage polkadot dress, is readying for the day's making. People all over the world meet Sabrina feeling they already know her, not because she's a TV star or her image is splashed across *People,* but because they've read her diary. Instead of maintaining an arm's length aura of celebrity, Harrison closes the gap by inviting people to participate in her "Art of Becoming Yourself" camps, helping others learn the slow and sometimes painful work of embracing our humanity — across the globe.

"As we have barreled deeper and deeper into a technological/result-driven 21st century," says Sabrina Ward Harrison, "now more than ever I believe we need to be brought home to the presence of our living."

This morning, she's laid tables with white glue, brushes, plastic cups full of primary colors and found objects from the previous afternoon's seaside collecting. As we gather on the couches in the pavilion, she pulls out a piece of twisted rope and speaks about following a line, trusting our story as it is unfolding. This morning we are beginning with a free write, but none of us know which way the day will go. Neither, really, does she. We open our journals and follow the lines she slowly reads aloud:

I remember the colors of . . .
There was a tiny . . .

It's a loose writing style found in Natalie Goldberg's *Writing Down the Bones.* As we follow the prompts, scenes from childhood rise from the pages: toes on the seashore, five-year-old hands making hospital beds for worms. It's a style Sabrina follows throughout her work.

Sabrina's confessional approach was drawn out by her first editor, Erica Benenate. "My editor would come and go through my journals and I remember thinking: 'What am I doing exposing this stuff?' Like [writing] about the back of my thighs. I was never someone confident like that. You would talk about this stuff with close friends, but you didn't read about it anywhere. It was a pre-blog world, and I was so young. What I was interested in was giving a window into someone's life in progress. We need to know we're not alone and something about the admitting and being messy was okay with me. I remember wondering where I would have been had Anaïs Nin and May Sarton not shared their journals."

Harrison's first book was published when she was only 23 and still in graphic design school. Since then, she has been sharing her life through a vivid interplay of artwork, photography, writing and video, revealing an authentic approach to life and art that has encouraged people all over the world to come to a place of honest expression and what she likes to call "true living."

> "As we have barreled deeper and deeper into a technological/result-driven 21st century, now more than ever I believe we need to be brought home to the presence of our living."
> — Sabrina Ward Harrison

Sabrina's big dream is to preserve some land where she can live, work and teach. "That's the pinnacle," she says. "That's what I would love, but I think if you are trying to support yourself and you have this big vision, it does take a village. There have been moments where I found myself asking: where do I go from here? But I believe this is when we need to trust the story, follow the adventure and believe that whether it comes together next year or 40 years from now, that it will come to be."

It's a faith her parents instilled in her. Her mother and father created an imaginative, free world for her and her younger sister. Her mom affixed swatches of fabric and sandpaper to the kitchen cupboards for tactile learning, and her dad built a yellow brick road in the backyard. She's been following that road ever since.

Her dad built a yellow brick road in the backyard. She's been following that road ever since.

It's our last night at the Highlands and we're trying to take a group picture dressed in vintage hats. Before we can snap a photo, Sabrina leads us in an after-dark frolic in the freshly cut grass. Nine streaming bodies careening through the thick night air.

True Living

Some might say that the way Sabrina Ward Harrison lives her life is unusual, not easily replicated. And yet, something in her speech, in her way of being, ignites a yes inside us, quiet as it might be.

Yes to freedom.

Yes to simpler.

Yes to together.

What can we expect to find when we choose to slow down, to stop comparing, to give expression to our feelings, put our hands to good work, to come closer? By reorienting our lives toward people, toward real life experiences and meaningful creation, we embolden ourselves. We say "yes" to our inner lives, to the deep spaces, to finding the hidden treasure in our life. Author Brené Brown would call it "daring greatly."

Before I left for this charming coastal home to work alongside Sabrina, I was filled with something unexpected, something I couldn't quite put my finger on. With my tongue uncharacteristically tied in knots, I finally spilled the truth to one of my best friends: *I felt nervous.*

I was nervous to go to camp.

It was the nervousness I felt before going to camp when I was eight. Except it was an adult kind of nervous, like I was fooling myself into believing that I could see all of the potential potholes ahead. I wasn't nervous the girls would tease me or the boys wouldn't think I was pretty. I wasn't worried I'd forget my bathing suit or that it would rain all week and we'd get stuck indoors playing Chinese checkers. No. I was worried because the trip, this camp, this four-day sojourn to meet a creative hero of mine, might not be all I hoped it would be.

I was making the arm's-length personal.

Something deep inside wanted me to fling myself into this trip with the unhindered expectation of a five-year-old. I wanted to believe with my younger heart (the better, freer, lighter heart) that it would be good. That it would be epic. That I'd hold it up like an Everest climb. And something deep inside was telling me it was true.

And yet my adult self warned me to be careful. To not care too much. To not get too excited. To set my expectations just a little bit lower.

And my five-year-old heart was telling my 31-year-old head to take a hike, to let it go, to believe with every inch of my being that it was possible. And somewhere in me, something was shouting *YES*.

The "yes" of my five-year-old, pre-period, pre-heart-smash, pre-career-detours, pre-fall-outs self.

And that's the problem with the real. It's a mess. There's a potential for hurt. So much hurt. And there is a potential for good. So much good. We just don't know which way it will go. But if we don't start planning, don't try leaping, then things get small. And online, it can feel like we're living in a tin can. Things are cramped. We need to climb out for air.

> What is the push of our age? Alone, complex, consume, faster. How do we push back? Together, simpler, sustainable, slow.

Power In The Particular

I've shared the story of our experience with Sabrina Ward Harrison because of its tangibility. It is the story of one group of people, in a particular place and time, and the relationships forged there.

Sabrina Ward Harrison's approach to work and relationship is deeply personal. It is tethered to time and place.

"Knowledge is concrete and particular," writes Canadian essayist and philosopher John Ralston Saul. Deeply offline connections to "people in particular," to quote *The Brothers Karamazov,* are our path to cultivating a meaningful existence in the 21st century. That means fewer screens with distant acquaintances and closer, deeper engagement with those we can see and touch.

The sad irony is that while our digital technologies have served the lonely by bringing them into relationship, whether it be through online dating chat rooms or webcast church services, many of us have grown *more* lonely as we continue to pursue many shallow connections rather than strengthening a few strong commitments. There is power in the particular. Localized work and relationships dwell in our lives in ways our online ones can only mimic.

And as Adrienne Rich points out, "the difference between a life lived actively, and a life of passive drifting and dispersal of energies, is an immense difference. Once we begin to feel committed to our lives, responsible to ourselves, we can never again be satisfied with the old, passive way."

In Wendell Berry's view, a new technology should not replace or disrupt anything good that already exists, including family and community relationships. Instead of asking ourselves if a new technology will make our work faster, easier, or more novel, he suggests exploring the answer to the question: *Do I wish to purchase a solution to a problem that I do not have?*

In her book *New: Understanding Our Need for Novelty and Change,* behavioral-science writer Winifred Gallagher takes Berry's question further, suggesting a series of considerations one should ponder before adopting something new. As we learned in Chapter 4, she draws on the expertise of psychologist Oshin Vartanian who advises us to consider the brain's limited energy and ask the following questions before engaging with something new:

- Why should I take this up if my daily scripts are doing a good job for me?
- Why exactly do I need another gadget?
- It will incur certain mental costs, so where will those resources come from?

These questions will serve us well in the brave new world ahead. In Gallagher's expert view, the best advice for living in a world with potentially limitless distractions is to be selective and to use moderation.

Artificial Intelligence and the Future of Humanity

We find ourselves in a remarkable time when the stuff of science fiction is coming to life all around us. "When it comes to computers" says Rodney Brooks, a robotics professor at the Massachusetts Institute of Technology, "who is us and who is them is going to become a different sort of question." We may be creating products, ultimately, to replace ourselves.

Indeed, this has been the arena of the Turing Test — an annual battle between the world's most advanced artificial intelligence (AI) programs and ordinary people — for decades. The test was introduced by Alan Turing in his 1950 paper "Computing Machinery and Intelligence," which opens with the words: "I propose to consider the question, 'Can machines think?'" Since that time, competitors have set out to discover whether a computer can act "more human" than a person.

Since 1991, the Turing Test has been administered by Hugh Loebner, a bankrolled former baron of roll-up disco dance floors. When asked his motives for orchestrating the annual competition, Loebner cited "laziness" to Brian Christian of *The Atlantic.* "His utopian future, apparently, is one in which unemployment rates are nearly 100 percent and virtually all of human endeavor and industry is outsourced to intelligent machines." (Source: *The Atlantic,* "Mind vs Machine," March 2011, p 58)

Fast forward to Spike Jonze's film *Her,* and you get the idea. In *Her,* Theodore, played by Joaquin Phoenix, falls in love with a computer program — a machine that can "think" — calling into question the limits of the body, computing power and human emotion.

High-tech entrepreneur Ray Kurzweil views this example not as fiction but as a rendering of the brave new world we are about to enter.

"A lot of the dramatic tension [in the film *Her*] is provided by the fact that Theodore's love interest does not have a body," Kurzweil writes in a review of the film. "But this is an unrealistic notion. It would be technically trivial in the future to provide her with a virtual visual presence to match her virtual auditory presence." He's talking about a future where he envisions something called the "singularity."

> Ray Kurzweil popularized idea of the Terminator-like moment he calls the "singularity" — when artificial intelligence overtakes human thinking. Kurzweil dreams of the moment that man and machine converge.

The singularity is a point at the center of a black hole where the laws of physics no longer make sense. Kurzweil raised the profile of the singularity concept in his 2005 book *The Singularity is Near,* in which he argues that the exponential pace of technological progress makes the emergence of smarter-than-human intelligence the future's only logical outcome.

In his role on the advisory board of The Singularity Institute for Artificial Intelligence (now, the Machine Intelligence Research Institute [MIRI]), Kurzweil and others work toward the Institute's vision of helping "humanity prepare for the moment when machine intelligence exceeds human intelligence." At first, the group operated primarily over the Internet, supported financially by transhumanists. (*Transhumanists,* as their label suggests, want to augment humanity, to extend humanity to a superior form with extra capacities beyond human limits, such as greater intelligence, healthier and longer life, even immortality.) The Singularity Institute has hosted numerous conferences in Silicon Valley, bringing together scientists and technologists to imagine a future where human and computers converge. If that happens, it will alter what it means to be human in ways that are impossible to predict. Today's artificial intelligence researchers have warned that it is imperative to develop ethical guidelines now, ensuring these advances help rather than harm.

> How much can we merge with technology and still remain human?

Ray Kurzweil is now Director of Engineering at Google. This is not fiction. He has been quoted in the *New York Times* as saying: "By 2029, computers will have emotional intelligence and be convincing as people. This implies that these are people with volition just like you and I, not just games that you turn on or off."

Everyone's allowed their theories. "It's just that Kurzweil's theories have a habit of coming true," says Carole Cadwalladr in *The Guardian.* Kurzweil predicted the explosion of the Internet when it was only being used by a few academics; he foresaw that by the year 2000 robotic leg prostheses would allow paraplegics to walk (the US military is currently trialing an "Iron Man" suit) and that "cybernetic chauffeurs" would be driving our cars (as they are already at Google). Kurzweil believes the future is near.

In the event of his declared death, Kurzweil will be preserved by Alcor Life Extension Foundation, perfused with cryoprotectants, vitrified in liquid nitrogen, and stored at a facility in the hope that future medical technology will be able to repair his tissues and revive him someday.

The Guardian reports that since Ray Kurzweil joined the company, "Google has gone on an unprecedented shopping spree and is in the throes of assembling what looks like the greatest artificial intelligence laboratory on Earth; a laboratory designed to feast upon a resource of a kind that the world has never seen before: truly massive data. Our data. From the minutiae of our lives."

Google has bought almost every machine-learning and robotics company it can find, including Boston Dynamics (which produces life-like military robots), as well as the secretive and cutting-edge British artificial intelligence startup DeepMind.

"The future," says Cadwalladr, "in ways we can't even begin to imagine, will be Google's."

Artificial Intelligence (AI) tools, technologies that exhibit the intelligence of machines or software, have been created to overcome what Kurzweil calls our "limitations," the limits of the body, of the mind, of mortality. What if we've got the trajectory wrong? What if the very limits of humanity are our map to finding meaning?

> What if it's these very human limitations that, instead of rising above, we should be drawing ever closer to?

Virtual Reality

Virtual reality (VR) is the natural extension of every major technology we use today. It is the ultra-immersive version of all these things — movies, TV, videoconferencing, the smartphone and the Web — that we'll continue to use to communicate, learn, entertain ourselves and escape.

The June 2014 cover of *Wired* magazine heralds the next great global shift since the iPad: Oculus Rift — the world's most immersive virtual reality goggles. Mark Zuckerberg gestured at the possibilities himself in a post in March when he announced that Facebook had purchased Oculus VF for $2 billion: "Imagine enjoying a courtside seat at a game, studying in a classroom of students and teachers all over the world, or consulting with a doctor face-to-face — just by putting on goggles in your home. That's the true promise of VR: going beyond the idea of immersion and achieving true presence — the feeling of actually existing in a virtual space."

That's because, by combining stereoscopic 3-D, 360-degree visuals and a wide field of vision — along with a supersize dose of engineering and software magic — Oculus has found a way to make a headset that does more than just hang a big screen in front of your face; it hacks your visual cortex. "As far as your brain is concerned," writes Peter Rubin for *Wired,* "there's no difference between experiencing something on the Rift and experiencing it in the real world."

Attached to the Real World

What happens when we are all immersed in our virtual realities the way we're swimming in the sea of our smartphones? If we are able to live inside our best dreams — like immersion virtual reality promises — why would we ever want to leave? Perhaps because we have needs to meet: a baby that needs feeding, a sick spouse to tend to, a friend to visit, or an aging father to care for.

To return to our central question posed by Albert Borgmann: *What happens when technology moves beyond lifting genuine burdens and starts freeing us from burdens that we should not want to be rid of?*

Instead of rejecting our limits, we must abide in them more fully, together holding firm to our humanity with the strength of a giant. It is time to turn the tide.

> "The hatred of the body and of the body's life in the natural world, always inherent in the technological revolution (and sometimes explicitly and vengefully so), is of concern to an artist because art, like sexual love, is of the body. Like sexual love, art is of the mind and spirit also, but it is made with the body and it appeals to the senses.
>
> "To reduce or shortcut the intimacy of the body's involvement in the making of a work of art (that is, of any artifact,

> anything made by art) inevitably risks reducing the work of art and the artist itself . . . I am not going to use a computer because I don't want to diminish or distort my bodily involvement in my work. I don't want to deny myself the *pleasure* of bodily involvement in my work."
>
> — Wendell Berry

When we make or fix something ourselves, we feel a fierce attachment to it, whether it be a lamp, a limb, or a marriage. Fixing is motivated work; it requires mental and bodily engagement as we connect to the particular. Repair Cafés and iRepair.com have emerged to feed the growing desire to take broken or damaged items and give them a new lease on life.

Begun by Martine Postma, in an effort to increase sustainability in her local community, the first Repair Café opened in Amsterdam in 2009. Since then The Repair Café Foundation, a Dutch non-profit, has supported growing numbers of "hackerspaces" around the world where people come together to fix items (vacuum cleaners, toasters, even old sweaters) that would normally have been tossed out. They are finding what Berry calls "the pleasure of bodily engagement" in their work.

Interestingly, the groups aren't entirely comprised of mechanical people; some are IT workers, but there are also sewers, gardeners, editors, and even anesthesiologists. Knowing how to make repairs is a skill quickly lost. Society doesn't always show much appreciation for people who retain practical knowledge, so they are often left standing on the sidelines. iRepair online tutorials, YouTube videos and Repair Café events create a space for valuable practical knowledge to be passed on, learned and used.

In the same way, something made for us by others — whether it be a pair of mittens, a piece of furniture or a home-cooked meal — holds a sensibility that bought things don't share. The investment of time, energy, thought and care are wrapped up in the yarn, the heartwood and the root vegetables.

It was a desire for this kind of investment that led journalist Andy Johnson to turn his hands to woodworking. Johnson grew up helping his dad, who was a painter, with home renovations. These early experiences resulted in an appreciation for not just the act of working with his hands and doing something tangible, but for the men and women who do it as a way of earning a living. As a result, growing up and into adulthood, Johnson was always making things, whether it be building a bookshelf, turning old doors into a hutch, or renovating his own house.

"I had to," said Johnson. "Although it was just a hobby, it was essential."

As a journalist working at newspapers and later in online news, his daily work didn't provide much opportunity for working with his hands. And after ten years, his job had become more about generating traffic, baiting social media clicks, or getting on top of the latest "trending" topic on Twitter. Johnson felt more desperately than ever the need to be doing something tangible with his 40+ hours per week.

"When I first got into journalism it was because I loved being out in the community, on the front lines reporting the news, telling stories, informing readers about important things. I felt like I was giving back and actively contributing to my community in an important way. As that changed and the job became more about watching the Internet for trends and then jumping on them, trying to capitalize on stories that were developing or already out there, it just started to feel soulless to me. Not what I wanted to be doing with my life."

Johnson traded in his byline for a ticket to California to study woodworking at College of the Redwoods, one of the most well-respected woodworking schools in the US. Following in the tradition of legendary cabinetmaker James Krenov, whose books had inspired Andy for years, Andy is learning about commitment to craftsmanship, the subtle use of hand tools and techniques, and woodworking basics such as dovetails and mortise-and-tenon work.

"We are becoming aware of the fact that much of our life is spent buying and discarding, and buying again, things that are not good," wrote James Krenov. "Some of us long to have at least something, somewhere, which will give us harmony and a sense of durability — I won't say permanence, but durability — things that, through the years, become more and more beautiful, things we can leave to our children."

It's this longevity, this power and meaning in permanence, that speaks to Johnson, who plans to open his own custom-order shop in Toronto.

"Woodworking is just so different from my work before. It's not about jumping on the latest trend and trying to cash in (at least for me it's not). It's about finding a way to do good work, to make good things, to share those things with people, to improve their quality of life and ultimately to make them happy. Most importantly, it's about building things well, so they can last for generations. So you only have to buy them once . . . There's something about using a handsaw, chisels and a plane to shape boards into a piece of furniture that is good for the soul . . . And though it may never be very lucrative, though I may struggle to make ends meet, it will be a happy struggle. It will be worth it."

> "Making your own plane, tuning it to sing like a fine instrument, and training your hand and arm to work in such a way that whisper-thin shavings fly from this piece of ancient technology . . . it leaves behind a surface that is smooth to the touch but somehow glows with handmade warmth. That is an awesome thing to experience."
>
> — Andy Johnson

Making is personal and powerful more than ever in a throwaway culture. "Time is precious," says Janine Vangool, editor, designer and publisher of *UPPERCASE,* an international magazine for the creative and curious. "Showing someone that you took time out to make something heartfelt is powerful."

Putting together a good old-fashioned made-by-hand gift for someone doesn't have to be something complicated, just something simple showing your loved one that you took some time. It could be a handmade card, some banana bread, a handwritten letter, a collage of pretty pictures, or a quirky little figurine — something from the past, packaged up with care.

Perhaps our greatest concern with "technological progress" should be the degradation and obsolescence of our bodies, suggests philosopher and farmer-poet Wendell Berry. "Let us see the good of the work, the good of the body, and ensure we do not, consciously or unconsciously, say goodbye to these things," writes Berry. "It is in being tethered to land, to people, to body that we find our meaning, and without them we are left adrift."

Where Do We Find Happiness?

The other evening, instead of texting my next door neighbor to ask a question, I bundled my two elder kids up and scurried across the front lawn in –15°C winds and knocked on the door. Instead of firing a couple of short messages back and forth (which seemed like better judgment once we were outside, freezing) my neighbor and I stood on the front porch talking about the challenges of parenthood. And then I saw it: the crack in her demeanor, tears at the ready. My neighbor was needy, and my physical presence let it come out.

"Weakness carries within it a secret power," writes Jean Vanier, founder of L'Arche, an international federation of communities for people with developmental disabilities and those who assist them. "The cry and the trust that flow from weakness can open up hearts. The one who is weaker can call forth powers of love in the one who is stronger."

Consider this scenario: the kind older gentleman on your block has had nothing to eat tonight. In fact, he's had little all week. His pension check got misrouted, and he hasn't had the wherewithal to get himself over to the food bank. Upon learning of this, you fill a bag of groceries, carry over a meal and some money to help him through. You see the need; it is at your front door, and you rise to help.

If people of character are formed by attachment, local cultures and local responsibilities, this suggests that our "up-at-all-cost" mentality may be the wrong target. Our technological advancements have one trajectory: *up. Up* the corporate ladder, *up* in social status, *up* and away from our surroundings, our physical demands, from the bore of the everyday.

We are headed up, yet the happiest people in the world are the Danes sitting *down* to have dinner.

> The happiest people in the world feel deeply committed to their families, neighbors and immediate community.

In 2013 the United Nations declared March 20 the International Day of Happiness, recognizing the relevance of "happiness and well-being as universal goals and aspirations in the lives of human beings around the world." In 2014, Denmark was listed as the happiest nation in the world. Apparently, one of the key factors in their happiness can be found in their spirit of "hygge," which translates as "coziness," but is really more of a complex sense of intimacy, community and contentment that generally comes with having meals with friends and family.

As much fun as it seems to read old classmates' vacation plans, the reality is that passive engagements don't truly make us content. "Happy people tend to be more spiritual and engage in more active leisure — things like dancing and joining a sports team," says Mark Holder, a professor of psychology at the University of British Columbia. "And happy people are less likely to engage in passive leisure, like being on the computer or watching TV."

While time spent alone can recharge you, research shows that happiness comes from more "pro-social" behavior. "Happy people tend to be energized by the social world and thrive on the company of others," says Holden, "even if it's an impromptu conversation at the grocery store." (Source: "Four Surprising Happiness Busters," *Chatelaine,* Feb. 2012)

Choose active life over passive hobbies.

The value of pro-social behavior is what Kirstine Stewart, now head of Twitter Canada, found during an afternoon of face-to-face conversations. In her role as a top executive at the CBC, she was slotted to participate in a Human Library event at the broadcaster's Toronto hub. Visitors to a Human Library are given the opportunity to speak informally with a varied group of "people on loan" — living, human books. In 15-minute slots, Stewart engaged one-on-one with a wide variety of individuals, tweeting at the end of the afternoon that she felt unexpectedly "rejuvenated."

Often, it feels like too big a hurdle to get out the door or, in Stewart's case, down the elevator. We expect to be sapped by in-person interactions when, compared to the same amount of time spent in front of our laptop, we experience the opposite. Broadcasting information, whether it be on TV, through social media, or by way of mass emails, does not allow for the give and take communication demands: that is, the exchanging and analyzing of ideas. These exchanges lead us closer to the Danish experience of "hygge."

Even "speaking [on the phone] is easy," says one executive interviewed by the *Harvard Business Review,* "but careful, thoughtful listening becomes very challenging. For the most important conversations, I see a real trend moving back to face-to-face."

Even in corporate America, it seems, the personal is making a comeback.

Mindfulness in the Age of Complexity

Novelist Flannery O'Connor once wrote: "You have to push as hard as the age that pushes you."

From Abraham Maslow's extraordinary understanding of human nature comes this gem: individuation requires resisting unhealthy enculturation.

What would we say is the big push of our age? To consume. More information, more products, more connections. How then, do we push back? To slow, be present, draw closer.

One way is to approach life with what Sabrina Ward Harrison calls *reverence.* Harvard psychologist Ellen Langer would call the same thing *mindfulness*: the process of actively noticing new things; being fully alive and living in the present.

> "It's going to sound corny, but I believe it fully: Life consists only of moments, nothing more than that. So if you make the moment matter, it all matters."
>
> — Ellen Langer

Over the past four decades, Langer's research on mindfulness has influenced thinking across a wide range of fields including behavioral economics and positive psychology. She makes the case that we can boost performance, ignite creativity and reduce stress by paying attention to what's going on around us, instead of operating on auto-pilot.

When you actively notice new things, "it puts you in the present," explains Langer. "It makes you more sensitive to context and perspective. It's the essence of engagement. And it's energy-begetting, not energy consuming."

This helps explain why, when Joshua Fields Millburn, co-founder of the popular blog The Minimalists, chose to limit his Internet access at home, he experienced it as the best decision he'd ever made. It was energy-begetting not simply because he had put the brakes on using the Internet at home, but because it was a product of conscious decision. The choice was made intentionally.

In living mindfully, we pay attention to whatever we are doing and "suck out all the marrow" — to use the wonderful phrase by Henry David Thoreau, who also said: "Wealth is the ability to fully experience life."

There are many advantages to mindfulness. We become more deeply absorbed in what we are doing. It's easier to pay attention. We remember more of what we've done and, as a result, we're more creative. We appreciate the people we are with. When you're mindful "you're able to take advantage of opportunities when they present themselves," explains Langer. "You avert the danger not yet arisen. You like people better, and people like you better, because you're less evaluative. You're more charismatic."

The charisma that Sabrina Ward Harrison exudes is not intrinsic; in Langer's view, it's been cultivated by thousands of mindful decisions. This means that we, too, can cultivate this way of being. We can train ourselves to be fully present, to live 100% in the present moment, and it begins with cultivating attitudes of reverence — remembering that every person has intrinsic value and is therefore worthy of our full attention.

> Every person has intrinsic value and is therefore worthy of our full attention.

A People-Focused Future

The on- and the off-line worlds are merging.

In Chapter 1, I spoke about good burdens — those we should not want to be rid of. Should we, as humans, ever want to be rid of the burden of people? Of the fallibility and hardship and mess of human relationships? What does the billion-dollar pursuit of artificial intelligence say about our humanity? What future will we choose through our daily scripts?

Ray Kurzweil's pursuit of immortality is nothing new, points out British philosopher John Gray. He argues in his book *The Immortalization Commission: Science and the Strange Quest to Cheat Death* that contemporary science is what magic was for ancient civilizations. "It gives a sense of hope for those who are willing to do almost anything in order to achieve eternal life." In his view, Kurzweil's Singularity is but one example of a trend which has almost always been present in the history of humankind.

In our pursuit of ease and control, drawn into the temptation to flee life's uncertainties, we are in danger of relinquishing our responsibility to each

> "My wish simply is to live my life as fully as I can. In both our work and our leisure, I think, we should be so employed. And in our time this means that we must save ourselves from the products that we are asked to buy, in order, ultimately, to replace ourselves."
>
> — Wendell Berry

other and our world. *Friends, we are in danger of writing each other out of our stories.* Jean Vanier tells us "we do not discover who we are, we do not reach true humanness, in a solitary state; we discover it through mutual dependency, in weakness, in learning through belonging."

Vulnerability is the cradle of the emotions and experiences we crave, says Dr. Brené Brown, author of the #1 *New York Times* bestselling book *The Gifts of Imperfection* (2010). It is "the birthplace of love, belonging, joy, courage, empathy, and creativity. It is the source of hope, empathy, accountability, and authenticity." Vulnerability is also the path few of us want to take. But if we want to find greater clarity in our purpose, deeper relationships, more meaningful spiritual lives and peace, *vulnerability is the path."*

Character, perhaps the key element missing in all our consumptions and pursuits, is found in the local, in the intimacy and accountability we foster face-to-face every day.

Chapter Challenge:

Write down the names of three people you miss:

____

____

____

Over the next week, contact them by phone. If possible, see them in person.

Now, set an intention to have a conversation — actual back-and-forth in-person or phone interaction — with one person (spouse, coworker, neighbor, stranger, grandfather, kid, friend) every day. It doesn't have to be long. Five minutes counts.

12

Little Eyes and Ears

Leading by Example

Make the Ordinary Come Alive
Do not ask your children to strive for extraordinary lives.
Such striving may seem admirable, but it is a way of foolishness.
Help them instead find the wonder and the marvel of an ordinary life.
Show them the joy of tasting tomatoes, apples and pears.
Show them how to cry when pets and people die.
Show them the infinite pleasure in the touch of a hand.
And make the ordinary come alive for them. The extraordinary will take care of itself.

— William Martin,
The Parent's Tao Te Ching: Ancient Advice for Modern Parents

If we are ever again to have a world fit and pleasant for little children, we are surely going to have to draw the line where it is not easily drawn. We are going to have to learn to give up things that we have learned (only in a few years, after all) to "need."

— Wendell Berry

When my daughter, Madeleine, began junior kindergarten two days after her fourth birthday, I had my reservations. As part of the curriculum for her age group (which included children as young as three) smartboards, iPads and TVs are incorporated into the classroom. On extreme cold and rainy days they watch movies in the gym over the lunch break (Smurfs, anyone?). In addition to this, one library block a week is dedicated to tech buddies. Here, kids learn to login to a computer with the help of a bigger kid, and they play "educational video games." These activities are slotted under the "media literacy" component of junior kindergarten.

The ages of three and four are a time for imagination and play, for learning conceptually and relationally. As a parent, I felt the media literacy component of junior kindergarten wasn't necessary. I contacted the principal of the school, a well-funded public institution on Toronto's west side, to discuss my concerns, particularly about the junior kindergarten "tech buddy" program; I didn't want Madeleine to participate, and I felt the school should reconsider that part of their curriculum.

I asked the school to change the 30-minute "tech buddy" time into a 30-minute "reading buddy" time, feeling the kindergarteners' time was better spent connecting to stories. It was a simple change and, I believed, a vital one.

My inquiries about iPads, smartboards, movies and video games were met with visible shrugs: "It's everywhere," they said. *Exactly,* I thought, *so why does it need to be here?* My requests were dismissed. The principal felt committed to the curriculum — a curriculum that was likely written before every infant in the world could login and navigate a tablet with their tiny thumbs.

I shared my story with Raffi Cavoukian, the famous children's troubadour and author of *Lightweb Darkweb: Three Reasons to Reform Social Media Be4 it Re-Forms Us.* "It is important for kids *not* to know how to log-in," Raffi practically shouted in response. "We need to reject this word 'balance' when it comes to computer technology and little kids. Balance in the early years is a red herring."

"When something is popular it doesn't mean that it's right. It's a very simple concept. There are billion-dollar tech companies racing to shape the future for everybody. They're making billions off of convincing boards of education that everyone needs a tablet. That's nonsense," says Raffi from his home on Salt Spring Island.

"Net evangelists cheer the virtual world with little reservation," he writes in *Lightweb Darkweb*. "Yet while there's scant evidence that daily online engagement contributes to, say, character development in our young, we do have evidence of Net dependence and [social media] addiction, with negative impacts on wellbeing and productivity."

In his book, Raffi introduces us to the work of Dr. Dimitri Christakis, a pediatrician and researcher at Seattle Children's Hospital. In an engaging TEDxRainier presentation, Dr. Christakis explains that from birth to the age of two, a newborn's brain triples in size — a growth spurt found only in this early time of life. What accounts for the growth are the synapses — the connections made within our brains and formed by our early childhood experiences. In fact, a newborn brain has 2,500 synapses [per neuron], and that number grows to 15,000 by age of three. Due to this rapid growth, Dr. Christakis is concerned with the technology children are encountering in those early months and years — particularly electronic visual media and its impacts on the neural connections made in our brains and on our capacity for human connection.

"We are technologizing childhood today in a way that is unprecedented," says Dr. Christakis. "In 1970, the average age at which children began to watch television regularly was 4 years . . . Today, based on research we've done, it's 4 months. It's not just how early they watch but how much they watch. The typical child, before the age of 5, is watching about four and a half hours of TV a day. That represents as much as 40 percent of their waking hours."

According to recent figures from Active Healthy Kids Canada, children and youth get an average of eight hours of screen time per day, while less than 20 percent get the amount of recommended physical activity. The American Academy of Pediatrics (AAP) Council on Communications and Media recommends families create a "media use plan" and set clear rules about TV, cell phones and other devices. That includes limiting kids' screen time to one or two hours per day, keeping children's rooms free of TV and Internet access, in addition to parents' modeling healthy consumption.

> By the time kids reach middle school, they spend less time with teachers and parents than they do with media.
>
> — Clinton and Steyer, "Is the Internet Hurting Children?"

Based on his research, Dr. Christakis has developed the "overstimulation hypothesis" which suggests that "prolonged exposure to this rapid image change [in TV programs] during this critical window of brain development would precondition the mind to expect high levels of input, and that would lead to inattention in later life." In other words, TV stimulation at the start of life means the chance these children will have problems paying attention later on increases dramatically. But, Raffi Cavoukian points out, there is good news about cognitive stimulation: Dr. Christakis reminds us that each hour parents spend reading to children, taking them to a museum, singing to them reduces the likelihood of inattention in later years.

Child development experts like David Elkind, author of *The Hurried Child,* tell us that the early years are a time for slow immersion, a season that will set the course for a child's life. When children have early access to the speed and gloss of tablets, smartphones and laptops — devices designed for adults — the pace of the natural world, with its muted tableau, can feel dissatisfying. Surely, as parents and educators, it is not our intention to breed this kind of discontentment in the early years.

When young children have access to the speed and gloss of tablets, smartphones and laptops, the pace of the natural world, with its muted tableau, can feel dissatisfying.

Published in 2013, *Lightweb Darkweb* urges parents to keep their underage (not old enough to know how to be online responsibly and with awareness) kids offline and to be conscious in their own consumption — all from the perspective from a tech enthusiast. (Yes, Raffi tweets.) Psychotherapist Andrea Nair agrees that when it comes to kids, (with the exception of some children with developmental needs) screens can wait, plain and simple.

The Hurried Child

The last thing we want to hurry along is childhood. It is the time when our moors are laid, emotional and psychic impressions made. A season for exponential growth, exploration and wonder. A time that, as every parent knows, is gone in a blink.

"Childhood has its own way of seeing, thinking and feeling," wrote French philosopher Jean-Jacques Rousseau, who is the principal architect of our modern understanding of childhood. To the young, the whole world is new, and children abandon themselves in their exploration. And, as anyone who has ever taken a walk with a two-year-old knows, this exploration cannot be rushed. Children learn through movement, and "movement helps orient and anchor [children] to their experience," says cognitive neuroscientist and educational psychologist Layne Kalbfleisch, "and it also helps with memory consolidation." (Source: CNN) And this movement tends to be slow.

In his book, *The Hurried Child,* David Elkind explains childhood with a plant metaphor. In his view, the late Jean Piaget offered the best-formulated version of a child's intellectual development as part of the ongoing larger process of biological adaptation. "For Piaget," he writes, "human intelligence is thus best understood as an extension of this adaptation. Thinking, like digestion, transforms incoming information in a way that is useful to the individual. But thinking, like vision, also adapts to the constraints of the surrounding world. Like a plant or an organism, thinking both changes and is changed by the environment."

The view of childrearing since the middle ages, in Elkind's view, has been to conceptualize children as growing persons in need of adult care and guidance. Today, a different view reigns: *child competence.* Child competence is the conception of children as competent to deal with and benefit from everything and anything that life has to offer. According to Elkind, this way of thinking seeped into our psyches as a rationalization for parents who continue to love their children but have neither the time, nor the energy, for childhood. As a result, children have access to information, technologies and experiences that they have neither the mental nor the emotional capacity for.

In many ways, we are rushing children. In our rapidly changing world, instead of shielding them, we are taking our children along for the ride. "The bewildering rapidity and profound extent of ongoing social change are the unique hallmarks of our era," says Elkind, "setting us apart from every previous society. For us, in the foreseeable future, nothing is permanent. Stress is an organism's reaction to this change, this impermanence. We live, therefore, in a time of wide-spread, deep-seated stress; it is a companion that is so constant, we may easily forget how completely stress pervades our lives."

And little ones feed on the tone we set, grazing all day on this stress, on our deep-seated emotions.

In the year 2000, a surprising study was published. Anxiety, historically an ailment among adults, had invaded the domain of childhood. The article in the American Psychological Association's *Journal of Personality and Social Psychology* told us that anxiety in children had jumped dramatically. The study found that as the cognitive demands of earlier schooling and media exposure increased, the social and behavioral demands followed, and it was making kids anxious.

Children have never been very good at listening to their elders, but they have never failed to imitate them. For better or worse, in word and in deed, our kids are following our lead. Mindfully approaching our own tech habits and setting clear expectations with our children, are the keys to joyful consumption, freeing us to enjoy the best of what Raffi calls the "lightweb," the good of our screen-based media.

My four-year-old tells me a show on Netflix is okay because "It's not scary, Mommy." Are we putting too much responsibility on children to make their own choices around the media and technology they consume?

"If we change the beginning of the story, we change the whole story."
— Raffi Cavoukian

Setting Expectations

When my neighbors, Phil and Kim, made plans to take their two boys on a holiday to Playa Mujeres, Mexico, they knew they wanted to park their work at home. Before leaving, they set out their commitments and expectations as a family.

"At home, we try (the operative word) to detach from technology during non-work hours and when the kids are awake. I have learned that with my older son it is effective to agree to rules in advance," explains Kim, a pharmaceutical rep and mother of two boys, age seven and one.

"In this case, we stuck to three rules for our holiday: First, be positive during the travel process. Cole isn't a fan of airports or line-ups. Now that he is seven, I have explained that flying is a part of traveling and he needs to remain positive. Second, healthy eating. On vacation we would be having a lot of treats. He would be allowed to eat what he wanted provided he balanced meals with fruit and vegetables. Third, technology: Phil, Cole and I agreed that vacation is about spending time with family and enjoying Mexico. While we would have devices, they would only be used in the morning before we left for breakfast and at the end of the day. Devices would not leave our rooms."

Kim and Phil did not do any work during the vacation; Kim left her work BlackBerry at home. The family felt they were able to enjoy meal times, memorable moments swimming and on the beach, and taking walks to the nearby marina to check out the lizards and other wildlife.

"I can tell you we did not have one argument about technology," says Kim, "or 'screens,' as Cole calls them. Agreeing to the rules beforehand was the key."

When asked if she thought the family engaged differently on holiday than if they had not set digital parameters, she replied with a hearty "yes."

"What we saw around us was very polarizing. Everywhere you looked, you saw parents and kids doing this," says Kim, swiping her finger from side to side. "On the beach, people were sitting there with their various smart devices swiping, swiping, swiping all day long."

Either families took Phil and Kim's family approach or the exact opposite, says Kim. "I saw parents of small children on iPads and other mobile devices all day long. Their children would play in the pool or on the beach while their parents sat there surfing the Web. I saw families that would come to meals with an iPad for each child, and the entire meal, the child would watch a movie. I saw children as young as 18 months being set up at lunch like this. It was difficult not to judge. Meal service could be slow at times, and I would send Cole to our pool chairs to draw or to the beach to dig (I could view him from the restaurant). I personally felt that was much more productive for him than giving him an iPad."

How can we do better as parents?

"That is a difficult question," says Kim. "Being a parent is hard. I think we need to set boundaries and then be accountable to them. I would never want my child to think that my phone is more important than him. Also, I want to teach my boys to have a healthy relationship with technology, and obviously my behavior will dictate that."

When it comes to young children, parents and educators are the media gatekeepers. Teen age can be a different story.

When it comes to teenagers' cell phones, says Ontario high school teacher, Jessalyn Hynds, "It's a part of them. If their phone is lost, they go into panic mode." There is even a term for it: *nomophobia,* as in no-mo(bile)-phone phobia.

"Now, I doubt we're going to see this phobia in the *DSM* anytime soon, but the anxiety revolving around smartphones (or lack thereof) is very real," says clinical psychiatrist Dale Archer. "We consider the smartphone an extension of ourselves, a best friend, even a soul mate. So the loss can be similar to losing a best friend."

This attachment to the smartphone is beginning at an early age. Teacher and mother of two, Marisa Ducklow, sees it developing both at home and in her high school classroom:

"I will say that as a mother I worry that my not-yet-three-year-old son knows how to navigate YouTube and that my one-year-old wants nothing more but to play with my phone. As a teacher, I see my students every single day literally stuck to their smartphones. When they think I'm not looking, they sneak a look at Facebook or quickly write a text. If there is a lull in class, they immediately plug in their headphones. They do not want me to do examples with them on the overhead because they just like looking at things on the digital projector . . . they would rather be passively observing some digitalized version of what I am explaining to them in real words and symbols that they can tangibly see and hear.

"It does worry me because I can see that their ability to concentrate has shortened from only a decade or so ago when I was in high school. It also worries me because they are addicted to the constant information being put out there for them to see."

It is also impacting their capacity for empathy.

Award-winning actress and theater teacher Astrid van Wieren has noticed a marked decrease in kids' and teens' ability to articulate and interpret facial expressions.

"Facial expressions, tone of voice and body language are all essential to communication and story-telling," says van Wieren, "and, as a performer, the tools of my trade. Over the past few years, as I've been teaching people in their late teens and early 20s, I've noticed a flattening of the voice, less expressive body gestures and muted facial expressions. It's been a subtle thing, but I feel it directly relates to texting and the online presence of young people's personalities."

"Teens seem quick to text LOL, but less likely to literally laugh out loud. With their heads down, they text furiously, sending out their communications online, but they almost seem to be wearing a neutral mask as they do it," she continues. "They'll add an emoticon, but seem less comfortable crying or laughing when acting in a scene. I'm not suggesting that young people are turning into robots, but when so much of their time is spent communicating their thoughts and feelings through their fingers, it seems they are losing true connection with the ability to read and interpret each other's physical emotional cues in the voice, face and body or to even convey clear emotional signals themselves."

Inherent in the dance of life is the ability to share the emotions of others; this is known as *mirroring.* According to Italian researcher Giacomo Rizzolatti, "Our survival depends on understanding the actions, intentions and emotions of others. Mirror neurons allow us to grasp the minds of others not through conceptual reasoning but through direct simulation — by feeling, not by thinking"

In mirroring, we establish rapport and empathy. Face-to-face, looking someone in the eye, we better understand what another person means, and we can ask for clarification when we don't.

Face-to-face, we experience "functional, moderate guilt," explains developmental psychologist Grazyna Kochanska. This feeling, he says, "may promote future altruism, personal responsibility, adaptive behavior in school, and harmonious, competent, and prosocial relationships with parents,

Dan Schawbel, author of the *New York Times* best-selling book *Promote Yourself: The New Rules for Career Success,* urges people to disconnect from technology every day. "In a past survey," he writes "we found that 40% of students feel like technology has hurt their soft skills, such as the ability to interact and build relationships face to face." These human connections, says Schawbel, lead to jobs.

teachers, and friends." (*Quiet*, by Susan Cain) These characteristics are especially important to foster in a time when a 2010 University of Michigan reveals that college students are 40 percent less empathetic than they were 30 years ago, with much of the drop having occurred since 2000. (The study, cited in Susan Cain's book *Quiet: The Power of Introverts in a World that Won't Stop Talking*, speculates that the decline in empathy is related to the prevalence of social media, reality TV, and "hyper-competitiveness.")

App developer Jeremy Vandehey remembers what he calls "the glory days" when we would spend hours playfully arguing with a sibling or spouse about some trivial fact. "My brother and I would spend entire afternoons having intense debates about the most Google-able, answerable topics," he reminisces. "The truth is, the answers never mattered as much as the conversation. It brought us closer. It taught us how to communicate. How to debate. Today, that intense argument would have fizzled out in two minutes, with Google having the final say."

> "The truth is, the answers never mattered as much as the conversation. It brought us closer."
>
> — Jeremy Vandehey

It's Complicated

In an interview with *Psychology Today*, danah boyd (she writes her name in all lowercase,) author of *It's Complicated: The Social Lives of Networked Teens*, draws on her work as a youth researcher and social media scholar, debunking some of the major myths about teenagers and their online habits.

Teens have less free time than ever before, and the time they do have, boyd suggests, is governed by adult involvement: school, extracurriculars, and being driven by their parents instead of walking or taking transit. Lunch hours are shorter, and curfews earlier, than they've ever been. Combined, these conditions remove the natural windows to socialize with friends; so, being humans and needing social networks, they have figured out how to get together online.

boyd has found that some teens spend more time with social media than they say they would like. They acknowledged being drawn into it and enjoying it so much that they lose track of time, and said it does cause some harm by subtracting from the time they can spend on other activities, including those that adults are encouraging them to do, such as schoolwork. But boyd is reluctant to call it addiction.

"But it is not clear that the harm outweighs the gains," says boyd. And, even if it does, she suggests the language of addiction is not helpful because it sensationalizes the problem. It implies pathology rather than a time management problem of the sort that all of us have to varying degrees.

"When I was in high school we had to dial-up to connect to the Internet and the noise it made was so loud my parents knew every time I was going online," says physics teacher Marisa Ducklow. "They got an Internet bill each month saying how many hours had been used, and if we went over I had some answers to give. Plus, even though we were connected to social sites like MSN Messenger and ICQ, our access was limited by the sheer physicality of accessing a computer. Today's kids face a different world."

For those carrying smartphones or tablets, choices have to be made hundreds, if not thousands of times a day. That requires a lot of brain power, emotional energy and restraint.

boyd points out that if we use the term "addiction" to refer to any activity that people enjoy and to which they devote great amounts of time, then "being 'addicted' to information and to people is part of the human condition: it arises from a healthy desire to be aware of surroundings and to connect to society." It's not the technology itself that draws young people in; it's the chance to communicate with peers and learn about their world. The computer is just a tool, like the telephone used to be.

> "Dear God, thank you that smartphones didn't exist when I was a teen. Sincerely, every mistake I made that didn't go viral."
> — Jon Acuff

"When adults see that children and teens are using computers and smartphones rather than playing outdoors or socializing in physical space, they find it easier to blame the computer and its supposed 'addictive' qualities than to blame themselves and the social conditions that have deprived young people of the freedom to congregate in physical places, away from interfering adults," says Peter Gray, Ph.D., research professor at Boston College.

> "Aside from the very serious problems of poverty and inequality, our nation's biggest offense against teenagers, and

> against younger children, too, is lack of trust," he continues. "Every time we snoop on them, every time we ban another activity 'for their own good,' every time we pass another law limiting their access to public places, we send the message, 'we don't trust you.' Trust promotes trustworthiness, and lack of trust can promote the opposite . . . They can't learn to trust themselves if we don't allow them to practice such trust."
>
> — Dr. Peter Gray

Free time, time for our minds to wander, time to engage in conversation, is essential to our development of self.

It's not only teens who are seeking escape from their overly orchestrated lives; Lego Group's anthropologists have learned that children play to escape restrictions and to work to hone a skill.

Free play is essential to children of all ages.

In a 2006 *TIME* cover article titled "Are Kids Too Wired for Their Own Good?" writer Claudia Wallis takes us into the lives of the Cox family, with their proliferation of screens. They are all in their three-bedroom home, but "psychologically each exists in his or her own little universe." And it is getting increasingly difficult to penetrate that universe says Dr. Karyn Gordon, a relationship and parenting expert for *Good Morning America.*

Gordon, one of North America's leading relationship experts and author of *Dr. Karyn's Guide to the Teen Years* believes that the number one guideline for parents is to teach our kids that technology is a privilege, not a right.

"We are their technology teachers."
— Dr. Karyn Gordon

"Overall, I'm a big fan of technology," says Gordon. "It's an amazing tool that when used responsibly, allows families to connect more efficiently and more often. I recently spoke to 500 high school students, and one teen said he tweets with his grandpa — his mentor — everyday! I don't think it should totally replace face-to-face communication; this is a skill that I recommend parents ensure they still teach their kids. The key is teaching our kids (and we are their 'technology teachers')."

Gordon loves using metaphors and believes that comparing technology to driving is a great illustration of this principle for kids.

"Driving a car is a privilege, not an automatic right," she says. "We need to be a certain age, be properly trained and follow the 'rules of the road,' or that privilege will be taken away. The difference here is that with driving, there is a test (written and road), a training manual, strict rules and speed limits that are all established for us. With technology, the 'rules' are left to the parents to decide and teach." Gordon suggests three simple rules of the road for the Web:

1. Use It Only for Positive or Neutral Comments
2. Talk Only To Those You Know
3. Do Random Check-Ups

With the third rule, Gordon recommends that parents let their kids know in advance that they will be reading their tweets, Facebook posts, etc. "This way, you are being honest with what you are doing (extremely important in a healthy parenting relationship) while making sure your kids are staying within the guidelines. I compare it to driving and speeding. We all know what the speed limits are (or least we ought to), but knowing that the police may show up at any time on the road helps to make sure we are staying within those boundaries. Knowing that these random check-ins may happen at any time will help our kids in their decision-making!"

And this education begins early on.

In a 2012 article "Is the Internet Hurting Children?" Chelsea Clinton and co-author Jim Steyer join in asking a vital question: What are the real effects of the Internet, and social media in particular, on children and teens? Clinton and Steyer conclude that "the promise of digital media to transform our lives in positive ways is enormous," but this optimism is predicated on a key condition: *If it is managed well.*

Let's Talk About It

The responsibility to teach children and teens to approach media with selectivity and moderation is up to us, and setting this example begins with conversation, explains Internet expert Sherry Turkle: "Conversations with each other are the way children learn to have conversations with themselves, and learn to be alone. Learning about solitude and being alone is the bedrock of early development, and you don't want your kids to miss out on that because you're pacifying them with a device."

According to a Stanford University study, more than ever, infants are missing out on parent talk, especially in lower-income homes, with long-term impacts on intellect. By the age of three, very poor kids hear some 30 million fewer words articulated aloud to them, and TV and Web-based voices, say experts, do not fill the void.

Babies are hearing some 30,000 fewer words articulated aloud to them, and TV and Web-based voices, say experts, do not fill the void.

With our older children, we need to enter more deeply in conversation, says Turkle.

In her *New York Times* article, "The Flight from Conversation," she writes, "connecting in sips doesn't work as well when it comes to understanding and knowing one another. In conversation, we tend to one another. (The word itself is kinetic; it's derived from words that mean to move, together.) We can attend to tone and nuance. In conversation, we are called upon to see things from another's point of view."

Instead of engaging our children in deep conversation, we are in danger of losing a new generation to numbing agents we've purchased and overuse ourselves. What are the long-term impacts (on focus, depth of understanding, reflection) of our kids' media consumption and what are they learning from our example?

Helen McGrath, a psychology professor at Deacon University in Melbourne, Australia, claims children are not as resilient as they once were because we are parenting with happiness, instead of independence, as our aim. Happiness, she says, should be the byproduct. When our kids are begging for the next episode of their favorite TV show, requesting a new iPod, or refusing to clean their room, this is a helpful long-term goal to keep in view. Pacifying our kids with screens doesn't build the resilience and self-control we are hoping to instill in our kids. It's also not helping encourage active play and time in the outdoors.

"We are now just beginning to understand that the growing disconnection between kids and the natural world is an increasingly serious social problem," writes Paul Cooper in his article "The End of Childhood."

Cooper explains the work of Dr. William Bird, a researcher with the Royal Society for the Protection of Birds in the United Kingdom. Bird has noted a steady increase in the diagnosis of childhood mental illness, such as ADD, and in the use of medication to treat it. But he also discovered evidence that simple exposure to nature has a huge effect. Cooper explains that "anything from unstructured play in a forest to a greening of the view from an urban classroom window — is an effective, non-pharmaceutical means of mitigating mental illness."

Children undertaking activities in nature appear to improve symptoms of ADHD (Attention-Deficit Hyperactivity Disorder) compared to urban outdoor activities or indoor environments, reports Dr. Bird (RSPB, 2007.)

"Children who survive through adolescence surrounded by gray walls and little time in the wilderness may not necessarily spend the rest of their lives believing that nature is a scary place, but the evidence suggests that their deficit of experience will result in an adulthood of generally higher stress and poorer health. Preserving and encouraging a natural environment is basic wisdom for the 21st century," says Bird. "An attractive future for humanity will be one in which all kids have the opportunity to roam, without fear, in an unspoiled land."

> "A child using his imagination to play a game in the woods isn't just having fun; he's setting a foundation for future independence, inner strength and an ability to resist stress that will last a lifetime."
>
> — Paul Cooper, *The End of Childhood*

Honest Toddler Says the World Is Magical

Bunmi Laditan, author of wildly successful blog and Twitter feed, The Honest Toddler, believes that seeing the world through children's eyes is magical, and we should give our kids as much opportunity for free play and exploration as possible:

> "Seeing the world through innocent eyes is magical. Experiencing winter and playing in the snow as a five-year-old is magical," says Laditan. "Getting lost in your toys on the floor of your family room is magical. Collecting rocks and keeping them in your pockets is magical. Walking with a branch is

magical. When we make life a grand production, our children become audience members and their appetite for entertainment grows. Are we creating a generation of people who cannot find the beauty in the mundane?

We constantly hear that children these days don't get enough exercise. Perhaps the most underused of all of their muscles is the imagination, as we seek desperately to find a recipe for something that already exists."

We can remember a time before the smartphone; kids can't.

As parents and educators, it is our great privilege to witness and provide space for this kind of wonder. And why is it essential that we ensure our children continue seeing through innocent eyes — encountering the magical? Because we are "the custodians of the pre-digital era," says Raffi. We can remember a time before the smartphone, they can't.

How To Miss A Childhood

In July 2010, special education teacher and mother Rachel Macy Stafford decided she was tired of losing track of what matters most in life, running from one task to the next with no room for rest, stillness or attentiveness. She began practicing simple strategies that enabled her to momentarily let go of distractions that had been sabotaging her happiness and preventing her from bonding with the people she loves most. Her blogging efforts at handsfreemama resulted in a *New York Times* bestselling book, *Hands Free Mama: A Guide to Putting Down the Phone, Burning the To-Do List, and Letting Go of Perfection to Grasp What Really Matters!*

"If technology is the new addiction," she writes "then multi-tasking is the new marching order. We check our email while cooking dinner, send a text while bathing the kids, and spend more time looking into electronic screens than into the eyes of our loved ones. With our never-ending to-do lists and jam-packed schedules, it's no wonder we're distracted. But this isn't the way it has to be."

For Stafford, this isn't just a multitasking or technology problem; it's the challenge of finding balance in a perfection-obsessed world. "It doesn't mean giving up all technology forever. It doesn't mean forgoing our jobs and responsibilities. What it does mean is seizing the little moments that life offers us to engage in real and meaningful interaction. It means looking our loved ones in the eye and giving them the gift of our undivided attention, leaving the laundry till later to dance with our kids . . . and living a present, authentic, and intentional life despite a world full of distractions."

It means not missing out on a childhood.

Parenthood, especially with very little children at home, is a mostly-drowning-and-sometimes-coming-up-for-air season. It can often feel like little more than survival. And we do what we can to survive.

For me, it's parking my two- and four-year-olds in front of the TV so they will be still and quiet for 15 minutes, while I put the baby down to bed. It's pulling out the iPad for humor and distraction when our middle child is sitting in the emergency ward waiting for stitches. It's using our iPhone to capture moments with photos or video because I don't have an extra hand to lug around my camera. It's checking in by FaceTime for a moment of connection when the kids are with a babysitter. It's about making life *work.*

But, in parenthood, just like the rest of my life, it can get out of hand. The 15 minutes turns into two hours, and the demands for more never seem to cease. Sometimes I have to "break" the television and reroute our efforts, redirect our attention. *Throw the cardboard boxes in the backyard and pull out the duct tape.* Other times I've got to make more of an effort to pull out the technology and connect with faraway family because I am not good at carving out time for that. I am figuring it out. And that's just it. We are all figuring this out, one blunder, one shining moment, one binge watch at a time.

Finding the Time

When it comes to our children, we want the fullness of it all. The depth of their eyes seeing into ours, the thrill on their faces when we watch them score a goal, the sorrow etched in the creases of their temples after a fall. We want to see it all, and they need to know we do — that they have us, all of us — and we need it, too. Because there are enough barriers in life, without the distraction of screens, our perpetual state of busyness, to separate us.

One of the big things I wanted to discover through my online fast was how my lack of Internet would impact the kids. By the end of 31 days, our eldest child was fiddling with my iPhone less, because I was. They are little mini-me's, our kids. They love to copy; it's how they learn. By removing the email access on my phone, I also removed its sense of urgency from my life. Also, by not accessing Netflix or YouTube, our TV watching options were significantly depleted. This meant decreased interest in screen time for our oldest, and more play for all of us. According to just about every study ever written about children and media consumption, this was good news.

Creating vs Consuming a Life

Author and educator Cecile Andrews, in her contribution to *Simpler Living, Compassionate Life,* writes: "In our culture . . . there is simply no room nor time for the spiritual life when you are preoccupied with getting ahead, making a profit . . . managing your investment portfolio, answering the phone, shopping or watching television."

By intentionally shelving things, we open up windows in our lives we wouldn't otherwise have. For me, stepping offline and forgoing the one-minute, three-minute, ten-minute smartphone check-ins gave me back more than an hour each day. That's an hour I could spend doing things I often lamented I had no time for. *I found time.* Here's the best news: you can, too.

One of my last phone calls with my grandmother took place during my fast. Hearing her thin Dutch voice was like a fresh coat of color on my white-walled heart. Her words to me that day were this: "Count your blessings one by one and thank God for all that he has done." As she spoke, I copied her words to a post-it and hung it on my fridge. Little did I know, those would be almost the last words she would ever speak to me. Making that call, which lasted no more than five minutes, the average length of a Facebook check-in on my phone, is a decision I will reflect on always.

"In rushing," Andrews continues, "we have no time for reflection, no time to notice what is going on around us. We can't reflect on warning signs that come to us . . . When we rush, we are much more likely to consume because we are ignoring the little voice asking if we really need this new thing." Andrews is confident that there is always a little voice speaking to us, telling us the right thing to do, but we ignore it because we have no time to listen.

Through my second-floor window, I stare out at a barren winter tree, its fingered limbs stretching high above the deadened foliage below. The ground is littered with twigs and dead branches, remnants from last night's heavy winds. In the ancient gospel of John there is a verse that reads: "He cuts off every branch in me that bears no fruit, while every branch that does bear fruit he prunes so that it will be even more fruitful." Like the tree in my front yard, there are branches in our lives that have long been dead. By cutting them off, what new life may grow?

Ultimately, moving us and our children away from complicated, high-speed entanglements — for many of us, dead branches — toward simpler pursuits, will lead us to greater peace and meaning. From this fruitful place, we are more free to create and confidently lead by example.

Chapter Challenge:

People, PARENTS: if you want more time in your life, join me in beginning to implement constraints around your Internet use. Start a 5 PM rule. Check email every other day. A self-described "CrackBerry addict"-mom friend of mine who formerly ran a multi-million dollar foundation, now checks email once a week. It can be done.

Whatever your rule: Do it. That nagging feeling that you are behind, that you don't have enough time to keep up with all of the emails, the reading . . . it will finally lift when you are no longer a slave to your inbox and newsfeeds. And your kids will be watching.

Get off. Try it for a week. It will change your life. *I promise.*

13

Making Space to Create

Discipline Is the Path to Freedom

But an artist's waiting . . . is not to be confused with laziness or passivity. It requires a high degree of attention, as when a diver is poised on the end of the springboard, not jumping but holding his or her muscles in sensitive balance for the right second. It is an active listening, keyed to hear the answer, alert to see whatever can be glimpsed when the vision or the words do come. It is a waiting for the birthing process to begin to move in its own organic time. It is necessary that the artist have this sense of timing, that he or she respect these periods of receptivity as part of the mystery of creativity and creation.

— Rollo May, *The Courage to Create*

Conversation is a catalyst for innovation.

— John Seely Brown, legendary innovator

I WANT TO SINK MY TEETH INTO THIS, I DO. But my exhausted chair-sitting self wants to let it go: to just keep sitting and surfing and being wiped out. I've been here for an hour spinning in circles, trying to dig into the writing when, in a fit of frustration, I snapped the lid of my laptop closed. First, I rifled through a pile of paper notes; then I picked up a book. As I began to read the pages aloud, I could feel the cogs in my head beginning to move forward. Quickly, *unsurprisingly* the fog cleared: *I could see the forest for the trees.* I was able to THINK.

There are 2.5 quintillion bits of information added to the Internet every day. As a result, each time we access the Web, we are offered something new, a shot of dopamine: a like! a share! an email! a purchase! Our egos are bolstered, our nervous energy absorbed. While ideas can spark online, it's more often through face-to-face conversations, sketches in our source books, extended hours lost in a project or even in sleep, that ideas grow legs.

Cultivating space, mentally and physically offline, is fundamental to our lifelong development as creators. Increasingly though, our space is mediated, and often cluttered, by the online demands of the Internet. As a means of necessity, textile designer and illustrator Samantha Cotterill restricts her time online. "While going on the Internet can offer a wonderful source of inspiration, its accessibility to all things creative can allow me to easily get lost in it and lose sight of how much physical work time is being neglected. I would mistakenly tell myself 'just 20 more minutes and then I'll start work,' and find myself still saying that three hours later. It takes time for me to get into a good work rhythm, and if I spend much of that time browsing the Internet, then another day will have gone by without any real work being accomplished."

> It takes every inch of my will to say "no." To close down the windows, the comments, the feeds and to place my attention here, on these words.

> Researchers at the Harvard Business School believe that always being plugged into the Internet can erode performance. One researcher observed that "certain cognitive processes happen when you step away from the frenetic responding to emails . . . Indeed, the history of science has shown us that insights not only occur in the laboratory but many [occur] while the scientist was engaged in a mundane task or even sleeping."

> Disconnecting gives us space to create, allowing us the mental real estate necessary for invention.

Like many artists, Cotterill turns to the Web for inspiration and validation but as her work has become increasingly digital — meaning she's spending more time in front of her computer — she's given her relationship with the Web a lot of thought.

"The types of relationships I have struck with the Internet are quite diverse, with each one occupying my life at varying degrees," says Cotterill. "Validation, Avoidance, Dependence, Healing, Experimentation, Pastime, and Survival are but a few examples of the types of relationships I can have with the Web."

After noticing the amount of wasteful time she was spending jumping back and forth between her work and the Web, Cotterill recently implemented a daily schedule and now has a clock that sits right in front of her. She has a set time when Web updates are made, and another set time for any networking that needs to be done for the growth of her business. She's even incorporated a period for free time, during which she can spend some good guilt-free time to have fun and just tinker about. (Her favorite hangout is the photo stream Flickr.)

"I am trying to bring back a balance that stops my ADD brain from wanting to quickly check something on the Internet just five seconds after opening up Photoshop to work. I have forced myself to only check emails two times a day, which has been much more difficult than I thought. I didn't realize how much I was going to my email, and even worse yet, checking my Flickr account to see how many new views I had since my last check three minutes prior."

When a friend suggested she get a trampoline to keep her Web compulsions in check, Samantha thought it might just be crazy enough to work.

"I now go and jump madly for a few seconds when I start noticing my little fingers starting to twitch as the 'need' to check things on the Internet gets stronger. I know it sounds silly, but it works. Trust me."

When asked what advice she would share regarding the interplay between the physical work of making and the online demands of the Internet, she says: "Make sure you are devoting the time you need physically and emotionally to create a good body of work, and set up a structured routine that will eliminate any wasteful 'let me just check this quickly' moments on the Internet."

As I dig into this, the 13th chapter of this book, I find myself wishing I too had a trampoline or no Internet access, or, you know, *more self-discipline.*

Paul Roden and Valerie Lueth run the Tugboat Printshop from Pittsburgh, Pennsylvania, where the husband and wife duo create beautiful wood-cut prints. When asked about their relationship with the Web they say, "It's inspiring to see all of the great stuff that people are doing and posting about online. But when we think about all the people making and doing and NOT posting about it every second, that's pretty mind boggling, too."

I often just don't have the willpower to keep myself from wasting time online. *Rescue Time* might help. It's an application that runs in the background of your computer recording how much time you spend doing different things, providing you with a little dashboard with a focus button that will stop you from running applications and websites it considers a distraction.

Focus. It can feel like a mythological creature in the digital age. But nothing is more central to the creative pursuit.

Longing and Commitment

"A common symptom of modern life is that there is no time for letting the impressions of the soul to sink in. Yet it is only when the world enters the heart that it can be made into a soul. The vessel in which soulmaking takes place is an inner container, scooped out by reflection and wonder."

— Thomas Moore, *Care of the Soul*

All creativity begins with longing to see something new in the world.

Children know this, for they are master wanters. They long for a tree fort in the backyard, even though the tree's circumference is a mere seven inches. They want a party with red streamers and an Easter egg hunt for 150 guests *this afternoon.* Their longing is strong. They can see what we cannot; their wants are deep and, in this way, they lead us, for creation must begin with desire.

We sketch out our longings in journals, think about them in the night hours, jot them on our iPads, post our growing interests on Pinterest pages. Slowly, we are giving name to our desires. And then, in the cause of soulmaking, we give them legs.

When we create, we give life to our longing; we are making what we want to see in the world.

For Jocelyn Durston, a longing emerged during her years in Africa, where she experienced the sure reward of a hard day's labor. Her longing was nurtured through the years spent working on her Master's in Environmental and Sustainable Development Economics. And it deepened whenever she watched the sun fall across the trees as day broke over the land. Her yearning eventually led her to abandon her meticulously designed apartment and cushy policy job in Ottawa to take up residency on a fledgling farm.

Durston was then a 31-year-old paper-pusher who used to dream of farming. One day, she tired of dreaming, bought a 1978 motorhome, sold and donated most of her belongings and moved to a small town (along with her cats, Fergus and Lola) to help some friends out and get her hands dirty.

"I'd been living in the city and I was starting to feel fidgety. I wanted to figure out a way to bring the farm aspect to my life," she recalls. "My friends Chris and Julie had been talking to me about coming out and helping with the farm they had started. I remembered their proposal, and so I emailed them. Within a week, it was a go."

Along with two couples and two toddlers, Durston began her "farm for a year" project. The goal was to transform 2.5 acres into a sustainable, flourishing, environmentally-sensitive working farm. Inspired by the belief that a healthy future requires the re-localization of food production, the small organic farm would serve as an experiment in gardening, caring for livestock and sustainable living. Using principles of traditional farming, with a respect for healthy living soil, the five endeavored to see just how much could be produced on a small plot of land by a bunch of novices.

"Today I get to be with my favorite people in the world and help them tackle their dream. When I leave my day job and get to the farm, it's like a weight lifts," says Durston, who chronicles her experience on her *Farm for Life* blog:

> "I'm pretty sure I can pinpoint when my childhood daydreams of 'Wouldn't it be cool to live on a farm?' became a more serious interest. I was 23 years old, in my third year of university, I was already interested in politics, food security, development, and environmental issues, but it was in the hours spent

> reading health food store propaganda while working quiet evening mall shifts that I really began to wrap my head around the global politics of food — and its connection to everything. After that, things just kind of started steamrolling."

The idea was born; and, then finally, she was committed. The next step was the Internet.

The Internet As a Source

"The Internet was invaluable throughout the whole process of the Farm for a Year/Farm for Life project," says Durston. "In all stages it was a valuable tool for research on how to do things. It was an even more valuable tool when it came to sharing our story through our website (blog), and through social media outlets (Facebook and Instagram, mostly). As a result of the following my blog got, we received numerous requests for interviews (print and radio), and we expanded our market reach (consumers sought out our farmers market stand after reading about our story on the blog). The blog and social media were also invaluable when it came to connecting with other individuals and groups who were like-minded and doing similar things as we were."

Even more than a practical source of information and market reach, the Web struck a deeper chord for Durston and her community, serving as a source of encouragement as they read stories online of others who were embarking on similar adventures.

"Without the Internet, I feel the Farm for Life project would have served as a source of inspiration and connection for far less people," continues Durston. "Our social circle, sense of community, and consumer base would have also been much smaller. There's so much interest in young farmers and local food sources right now, that harnessing the power of the Internet to take advantage of that, for me, is a no brainer."

> "Tools and technology, when subordinated to our desire for patience, self-reliance, skill and fortitude, are extremely helpful, even liberating."
>
> — Aiden Enns

A few years ago, in an incredible act of collaboration, gamers succeeded where scientists could not. Online players from across the world worked together to discover the structure of a retrovirus enzyme whose configuration had stumped scientists for over a decade. Gamers figured it out in 30 days. This is the first known instance in which gamers solved a long-standing scientific problem. The discovery was facilitated by Foldit, a website that engages the public in scientific discovery by using games to solve hard problems that can't be solved by either people or computers alone.

Janine Vangool, creator of the international quarterly magazine *UPPERCASE,* mines the Web for the best available tools. She particularly loves Evernote (think: an online corkboard with design and notetaking capabilities) and she uses it in virtually every facet of her creative life. "If I'm feeling out of ideas," she says in an Instagram video, "I go to Evernote where I've been saving stuff that I like for years, and invariably I will find something in there that will spark a new idea."

For Vangool, the Internet is first and foremost a tool. It has to be. As publisher/illustrator/editor she is responsible for assigning, editing and designing every issue of *UPPERCASE,* (which was nominated for magazine of the year at Canada's National Magazine Awards) as well as keeping up with an expanding subscriber base — and her preschool-aged son. She doesn't have time for surfing — the literal or figurative kind.

The ideas and resources found online can assist in nearly every aspect of our work, providing us with the raw goods to work with. It is up to us to finish the work.

While a fan of all things creative, Web-based or otherwise, Vangool believes it is important for people to cultivate a trust in their own instincts. "These days, it is too easy to get bogged down in the perceived perfection of Pinterest and the tyranny of step-by-step craft instructions," she says. She encourages her readers to unplug from these distractions from time-to-time. "Comparing yourself to others and following directions can be so detrimental to genuine creativity. Use your own ideas, your own resources, your own ingenuity . . . you will make something that is from you and your heart."

Every day, millions seek inspiration on the Web. One unique Flickr group called the Work-In-Progress Society (WIPS), celebrates the beauty in unfinishedness; members post photos of their works in progress and the community cheers one another on to completion.

Our communities, wherever we find them, can work like bricklayers, solidifying the foundation of our ideas and pouring mortar in the cracks as we build upward. Their affirmations fill in the gaps where otherwise, alone, we might falter.

"I am amazed and totally in awe of the Internet," says Karen Ruane, a UK-based embroidery artist, who is a member of WIPS. "It opens up so many possibilities for communication, interaction and learning, and I wonder constantly how we ever managed without it."

The Internet As Livelihood

For Karen Ruane, whose sophisticated white-on-white embroidery work is both a vocation and an obsession, the Web is the key to her livelihood. In addition to exhibiting her work across England, she offers online courses in embellishment and buttons to individuals across the globe, who take her classes from the comfort of their own homes.

"I try to make the Internet work for me, yet not take over," she says. "I don't want to be an administrator; I want to be a maker and a teacher. It is a conscious effort daily to set aside the time for both as separate aspects of what I do: embroiderer and online creator. I love that the Internet gives me an opportunity to reach the world — for free — in order to promote my work. I love that it allows me to teach in places like the US, Canada, Australia and Europe without leaving the house. Isn't that amazing? Having access to the Internet also allows me to keep up to date with contemporary art, seeing what is new and developing in terms of my peers."

Thousands of artists around the world make a living online. By selling direct through sites like Etsy, and crowdfunding ventures such as Kickstarter, Indiegogo and Australia's Pozible, the Internet has helped make creative enterprises economically feasible. That's the thing about marketing in the digital world: the opportunities for entrepreneurs to make a big splash, even without a marketing budget, are as endless as cyberspace itself.

Kevin Kelly, founding executive editor of *Wired* magazine, wrote an article titled "1,000 True Fans," based on the idea that a dedicated artist could cultivate 1,000 True Fans using new technology; the direct support of that number of fans allows an artist to make an honest living.

A True Fan is defined as someone who will purchase anything and everything you produce. Kelly writes, "They will drive 200 miles to see you sing.

They will buy the super deluxe re-issued hi-res box set of your stuff even though they have the low-res version. They have a Google Alert set for your name. They bookmark the eBay page where your out-of-print editions show up. They come to your openings. They have you sign their copies. They buy the t-shirt, and the mug, and the hat. They can't wait till you issue your next work. They are True Fans."

> "A creator" such as an artist, musician, photographer, craftsperson, performer, animator, designer, videomaker, or author — in other words, anyone producing works of art — needs to acquire only 1,000 True Fans to make a living."
>
> — Kevin Kelly

Assuming that True Fans will each spend one day's wages per year in support of the artist's work; that might be $100 per person, which adds up to $100,000 per year, which is a very good living for most people, particularly artists. It's the digital technologies of connection and small-time manufacturing that make this circle possible. The fan base can be cultivated, says Kelly, allowing hobbyists to turn their passion into livelihood. This is the Web at its best.

Completion: Discipline Is the Path to Freedom

Members of WIPS agree that while the Web is a substantive source of inspiration and resources, it's the creating and the discipline needed to complete the work that leads to a sense of freedom and accomplishment. They suggest that before you approach the keyboard, it's important to have a task in mind. The best way to treat the Internet is like an X-acto knife. Take it out of the toolbox, get the job done, then tuck it away for next time.

> Need help completing a project or to-do list — ending the day with a sense of accomplishment? Write a short list and work hard to get it done. Otherwise, there's an aptly-named app for that: Finish.

It's better to get lost in the making than lost in the Web.

Chapter Challenge:

In order to create, we need to set apart time to dream, to long and then to create. Perhaps longing must come first, in the quiet of Internet-free spaces, and then, after we've felt and understood a little more what it is we want, we can use the Internet as a tool to inform, equip and inspire that longing and creativity.

Begin by asking yourself these questions:

What do I long for?

____

____

____

What can I create out of that longing?

____

____

____

How can the Internet be a wisely used tool for me in creating this?

____

____

____

What limits will I give myself while using the Internet as a creative resource?

____

____

____

14

Hereon In

Check In, Check Out

Lasting change happens when people see for themselves that a different way of living is more fulfilling than their present one.

— Eknath Easwaran

Everything good I've ever gotten in life, I only got because I gave something else up.

— Elizabeth Gilbert, author of *Eat. Pray. Love.*

A PAINTER TAKES A PAUSE BEFORE HE STARTS A STROKE, hand hovering inches from the sheen of canvas, as if in prayer. And then the lines begin, the black of oil unfolding an inner blueprint. The outline comes quickly; it's a matter of minutes before he shuffles a few feet back to take in the full view.

It's the after-work that's painstaking. The palette, shadows and spaces for light. The smudging, resin, thickened layers giving the subtle sense of an ocean's tide. And in between, the mistakes. The lines unintended, the colors not right; but with them he moves onward, drawing them into the finished work. And, at some point, he says: It is finished. When, of course, it never is.

The metaphor of composing, and with it the improvisation essential to a painter or musician, is a proper one for life. "It allows for variety, mistakes, and change over time," writes Trudelle Thomas in her book *Spirituality in the Mother Zone.* It allows for "forgiveness and fluidity."

None of us has been a bystander in the digital age. The seas have swept us all, caught in a surge of information and entertainment too good to pass up. And now, here we are washed ashore, a bit ashamed, unsure of our footing as we rise to walk on altogether different terrain.

The Power of Habit

"All our life, so far as it has definite form, is but a mass of habits," William James wrote in 1892. Most of the choices we make every day may feel like the products of well-considered decision-making, but they're not. "They're habits," writes Charles Duhigg in his *New York Times* best-selling book, *The Power of Habit.* "And though each habit means relatively little on its own, over time . . . they have enormous impacts on our health, productivity, financial security, and happiness."

It's only over the past 20 years that scientists and marketing experts have begun to really understand how habits are formed and how they can be changed. As we take on a new set of habits, says Duhigg, one set of neurological patterns — our old habits — are overridden by new ones. The neural activity of our old behaviors are crowded out by new patterns. As our habits change, so do our brains.

We might already know about this, yet still we reach for the phone in our pocket 20 times a day. The good news is, the habits we have — even compulsive habits we have formed online — can be redirected. We can override them with new habits.

> "A habit is a choice that we make deliberately at some point, and then stop thinking about but continue doing, often every day. It's a natural consequence of our neurology."
>
> — Charles Duhigg, *The Power of Habit*

But we must begin with *one.*

In his chapter, "Why We Do What We Do," Duhigg recounts the story of Lisa Allen, a 34-year-old woman who had started drinking and smoking when she was 16 and had struggled with obesity and holding down a job for most of her life. When Lisa's husband left her for another woman, she flew to Cairo on a whim and there, jet-lagged in her hotel room, she accidentally lit up the end of a pen and tried to take a drag.

Lying in bed, she broke down. "I felt like everything I had ever wanted had crumbled. I couldn't even smoke right." This moment touched off a transformation.

"I need a goal," she thought. So, she chose one. To come back to Egypt and trek through an unnamed desert. "It was a crazy idea, Lisa knew. She was out of shape, overweight, with no money in the bank. She didn't even know . . . if such a trip was possible. None of this mattered, though. She needed something to focus on," writes Duhigg. "And to survive such an expedition, she was certain she would have to make sacrifices. In particular, she would need to quit smoking."

It was Lisa's focus on changing one habit that set off the shift, helping her reprogram the other patterns in her life as well. She was changing a *keystone habit.*

The human brain is very adaptive. If you repeat a behavior enough times, synaptic pathways get worn in. "It takes 21 days to form and break a habit," is the standard thinking. But new pathways aren't forged in an exact number of days. Everyone's brain is different, and habit formation relies on a variety of aspects including life experience and personality. And it's universally true that it's easier to start a habit than break it.

The one thing we know for sure? We have to start somewhere.

> "Whoever loves discipline loves knowledge, but he who hates reproof is foolish."
>
> — Proverbs

> "Those who had many habits dragging them down . . . each decided to focus upon just one 'keystone' habit. When they changed the keystone habit, other changes followed."
>
> — Charles Duhigg, *The Power of Habit*

Renounce to Gain

To proclaim technologies such as tablets, smartphones and television as *neutral tools* flies in the face of how people behave.

"Why do 90% of all families or households watch television after dinner? Is it because they decided that that's the best way to spend their time?" asks philosopher Albert Borgmann. "No, something else must be going on. And what's going on is that the culture around us — including work that is draining, food that's easily available, and television shows made as attractive as some of the best minds in our country can make them — encourages us to plop down in front of [the screen] and spend two hours there."

And the hours only continue to increase.

To quote Flannery O'Connor again: "You must push as hard as the age that pushes you." If we want a different, richer quality of life, then we have to push as hard against the pull of the most desirable technologies available to us. But most of us simply don't have that much discipline. I know I didn't. It's why I had to step offline for 31 days. And I *still* don't have the discipline I'd like, even after my extreme act of ditching the Internet.

There is a certain starvation of spirit caused by conflicting commitments, shortage of time and demanding work. We need enough imagination to envision a life beyond this labyrinth, to break our habits. Like Lisa Allen and her trip to Egypt, we must *will* something new.

But, as Ronald Rolheiser points out in his book *The Holy Longing,* most of us want the best of both worlds:

> "We want to serve the poor and have a simple lifestyle, but we also want all the comforts of the rich; we want to have the depth afforded by solitude, but we also do not want to miss anything; we want to pray, but we also want to watch television, read, talk to friends, and go out. Small wonder life is so often a trying enterprise and we are often tired and pathologically overextended. Medieval philosophy had a dictum that said: Every choice is a thousand renunciations. To choose one thing is to turn one's back on many others."

When we choose to simply dwell in a moment, not capturing it digitally or sharing it with an audience, we are choosing to trust our own memory. When we take a thousand selfies, we are choosing to discard our humility and modesty. When we log off for a weekend, we choose to ignore a million bits of news.

Each time we choose reality, we turn our backs on virtual reality. In choosing to visit the lonely, we say "no" to the need for robotic replacements. People or Pinterest. Humans or Hoodoo. Friends or Facebook. Engagement or escape. Marriages or machines.

Every choice is a renunciation.

The moment that changes everything is the moment we wake up to the truth that we have to give up something to accomplish our goal. We must renounce to gain.

> "Memory is not just the imprint of the past time upon us; it is the keeper of what is meaningful for our deepest hopes and fears."
>
> — Rollo May
> *Man's Search for Himself,* 1953, (p 220)

I confess that I can click around on the Internet with the best of them. I do it a lot — even as I write this book. And then there I am, an hour later, with nothing to show for it — and I'm paying someone to take care of my kids! And *I am supposed to be good at this stuff.*

I reinstall the Facebook app on my iPhone so I can share some pictures with my far-flung family and friends, and seven days later I catch myself checking my newsfeed for the 17th time that day, and I have no choice. *I must delete* or perish at the hands of Zuckerberg. And each time I do, I realize how little I need it.

"Discipline is the path to freedom," writes Anne Lamott. "This is another thing I hate, as I am drawn to sloth and over-consumption, and squandering whatever time I have left. But it is true, I promise. Discipline frees our spirits."

Your Detox: Where to Begin

How do we stop procrastinating and start implementing? Psychologists have no shortages of suggestions, but here are four to get us started. Over time, we can create space for simple, but energizing habits.

1. Get a Good Why
2. Start Easy
3. Break It Down
4. Be Gentle With Yourself

Go to the end of Chapters 7 and 12 and reread your Chapter Challenge answers. *What are you seeking? What are you longing for?* Sit with these thoughts for a moment.

Now, answer: What is most important to me?

(Ex: Time at home with family. Rest. More engaged in community. Creative hobby.)

____

____

____

What are the imposters that threaten or supersede the things that are important to me? (Ex. Nurturing your public persona. Lethargy. Distraction. Online obsession with your ex.)

____

____

____

1. Get a Good Why.

My why? To be a present person, to clear my head. I wanted a different quality of life, not a voyeuristic one. I longed for focus and quiet. I wanted to figure out the kind of person and parent I could and would be offline.

Now, write your why:

____

____

2. Start Easy.

Week 1.

<u>Pick a few things to do.</u>

This week, before you grab your smartphone or sit down at your computer, take a minute to jot down the things you intend to do. Here's mine:

- Email Sarah
- Check Twitter for ten mins
- Send one-page proposal to boss

Finish the list, and log off. Rinse, repeat.

Like every tap-happy iPhone user, I must train myself to focus. When I'm writing, I open a clean new document, go to the view tab and make it full screen. I have to log out of Facebook — it's just that little step that keeps me from looking at it. And then, if I'm lucky, I can focus for a good hour. But if I were smarter, I'd close the lid on the laptop and start writing on paper. Or I'd kill the Internet at home. Or I'd get Wunderlist (a sole-focus task management app). I'm getting there.

When we break things down in achievable pieces, we gain confidence and hope. Financial coaches like Dave Ramsey will tell you that when you have multiple debts, you should begin with the smallest, regardless of interest levels. Early in our marriage, my husband and I followed this advice with astounding results. When we paid off one of my husband's smaller school debts, we high-five'd each other. *Yes! We can do this!* Then, we were onto the next, and, finally, we paid off the big kahuna. Over the course of a year, by scrimping, saving and working our tails off, we did it, together. We were forming new habits of saving, and we had momentum.

> When we start easy, we achieve our goals, we gain momentum. *We are on our way.*

3. Break It Down.

Week 2.

<u>Simplify</u>

Plan how you are going to disentangle from the Web. What apps will you delete, what habits will you change, what are going to do with your time away?

Do use social media, don't live it. It can be a fantastic tool to give you direct access to a person you otherwise wouldn't or couldn't communicate with. FaceTime and photo-sharing are wonderful ways to bridge distances. Electric underpants to "feel" your partner a thousand miles away — perhaps not.

> We must arrest the time for ourselves. This may mean deleting certain apps from your tablet, creating an after 5 PM rule where you ignore your phone until morning, or creating an even more dramatic change — like only checking email once per day or week.

- ❑ Make it inconvenient to use apps you don't usually use. Put them on the last screen on your iPhone or tuck them away in folders. This removes the clutter and opportunity for distraction. Delete all unnecessary apps (try to get your list down to 20).
- ❑ Unsubscribe from email clutter. Deleting these every day is a time drain.
- ❑ Turn off all push notifications. These nasty notifications distract you from work, hobbies, people and passions. Keep your focus.
- ❑ Delete your social media apps. (If you want to post pictures, take the extra step and upload them to your computer, then post.) This will be the hardest and the most fruitful change of all.
- ❑ Look for websites, apps and organizations that make your life better, allowing you to access information and connections without sucking you dry. Wikipedia and Happify come to mind. Support these sites financially, if you can.

Week 3.

Speak to the Known.

How many times in a day do we speak to the unknown? Every time we publicly post anything on the Internet. Sure, we know a handful of people who will likely read it. But nothing is guaranteed. The smaller, more direct the audience, the freer you can be to speak — to really speak — to your intended audience. Meaning dwells in limitation.

It is a fundamental human need to form and maintain a number of lasting, positive and significant interpersonal relationships, say psychologists Roy Baumeister and Mark Leary. A lack of deep belongingness, they say, causes various undesirable effects, including a decrease in the levels of health and happiness, even higher levels of mental and physical illness.

Satisfying this need requires frequent, positive interactions with the same individuals and engaging in these interactions within a framework of long-term, stable care and concern. Despite the lure and excitement of new engagements, the need for some stable, caring interactions with a limited number of people is a greater imperative.

This week is about experimenting. *Quit your favorite social media site for one week.* Fill this time with get-togethers (one, minimum,) hobbies and phone/Skype calls (three, minimum) with those who love you most.

- ❑ I will quit ______________________________
- ❑ I will have coffee with ______________________________
- ❑ I will call:

1. ______________________________
2. ______________________________
3. ______________________________

- ❑ I will do:

____

Promote Work-Life Balance.

Since the 1990s, we have witnessed and experienced enormous leaps in communication technology, enabling us to stay connected with distant friends and family via email, Skype and FaceTime. These same technologies, however, make it more difficult to disconnect from our work, even hours after our shift ends.

"The danger is, when you get good at something, you get consumed by it," says Alan Kearns, founder of CareerJoy, a Canadian career coaching organization. The solution, according to Kearns, is to get good at other things: take up a hobby with your kids, volunteer for a cause you believe in, or join a sports team. Having many things that are important to us helps to round us out as individuals and keeps us from obsessing over a single area in our lives, says Kearns.

Week 4.

Cultivating Play and Rest.

> "If we want to live a wholehearted life, we have to become intentional about cultivating rest and play, and we must let go of exhaustion as a status symbol and productivity as self-worth."
>
> — Dr. Brené Brown

- ❑ Decide when you will check work email. Set notifications on your computer or phone. Talk to your boss about it and let colleagues know.
- ❑ Choose a checkout time. Whether it's 5:30 PM or 9, stick to it. Charge your phone *outside* the bedroom. Better yet, park it by the front door.
- ❑ *The People Prerogative.* Silence and put away your handheld device when you are talking to someone in person. The person in front of you is most important in that moment. Make them feel that way, and they'll do the same for you.
- ❑ *The Door Drop.* When you get home from work, drop your phone at the door with your keys. If you have a family, you will be more present to them and they will love you for it. If you are single, you will carve out a quiet haven at the end of your day, and you will love you for it.
- ❑ *Honor the Holy Hours.* You have a window of time when you first wake up that will set the course for your entire day. Don't fill it by checking Facebook. Instead, read a holy scripture, meditate on a goal, or simply sit in quiet. The time is yours and it is currency; don't spend it in the wrong place.

Week 5.

Set aside a Sabbath.

Begin setting apart a tech-free day: 24 hours each week where you disconnect completely. Power your phone all of the way down. Not on airplane mode. OFF. Tuck away your laptop. This day is going to be more liberating than your 16th check-in on Twitter, I promise.

Continue this practice for the rest of your days.

Week 6.

Plan your own fast from the Internet.

I suggest experiencing an Internet fast first hand (and then practicing a weekly tech-free Sabbath.) Begin by choosing a time frame. Make it as long as possible, no less than two days. Holidays, Christmas and long weekends provide good natural breaks. Write it on the calendar. Commit.

After Baratunde Thurston's unplugged vacation, he offered this checklist for those wanting to do the same:

- SCHEDULE A VACATION. Figure out when you can take a real break. If you want a true digital detox, two weeks is far better than one.
- ALERT YOUR KEY COLLEAGUES. A month before you leave, make sure that your key coworkers know that you'll be truly unavailable. This gives you time to work out any real problems your absence may create. As danah boyd says, "Warnings are the key to happy relationship maintenance."
- WARN EVERYONE. A week before D-day, send an email to a list of those who communicate with you on anything more than an occasional basis, alerting them to your departure. Make it clear to them that this is serious — no one will believe you're really capable of ditching the digital life.
- WARN EVERYONE — AGAIN! The morning of D-day, send an email to that list again. Make it emphatic — mine began, "I Have Left the Internet." If they don't understand that you're for real now, they can't be helped. They have, after all, been warned.
- TURN OFF AUTOMATIC SYNC ON YOUR PHONE. You can live without notifications from ESPN, Boing Boing and Mafia Wars for a few days — a couple of weeks even!
- SET YOUR AWAY MESSAGE FOR EMAIL. Your note should be courteous but firm: You will return no emails (though you may choose to leave emergency contact info).
- MANAGE SOCIAL NETWORKS. You can't really turn off Facebook, Google+, Instagram, and so on. So use your home page to establish your absence. Take a photo of a stark message like: I Won't Be Here Until [date of your return]. Use that as your profile photo.
- ESTABLISH YOUR EMERGENCY EXCEPTIONS. There must be some way for people to reach you. Set up a clear system with someone you trust, who can have access to your email and social media.
- TAKE A DEEP BREATH. Vacate. Completely. It'll be scary for a day or two. And then it will be great.

Disconnecting, unplugging, digital detox — call it what you like — these are just the beginning of forming new habits. What has the greatest impact is the decision to value and dedicate our attention to the people right in front of us.

> Hello, Mark Zuckerberg. The word is out. There is real, rich beautiful life afoot. You can help us find it. Shut Facebook down for a single day and let the world roll. We'll tell you the next day what we found.
>
> #digitaldetoxday#life
> Signed,
> Humans of the World

Slow Adoption: The Next Big Thing May Be No Thing at All

An Amish man told Kevin Kelly, founding executive of *Wired*, that the problem with phones, pagers and PDAs was that "you got messages rather than conversations." "That's about as an accurate summation of our times as any," quips Kelly on his blog The Technium.

Kelly's impression is that the Amish are living about 50 years behind us. Contrary to popular belief, the Amish *are* adopting new technologies — but at a much slower pace than the rest of us. "By that time," says Kelly, "the benefits and costs are clear, the technology stable, and it is cheap." Where our default to new technology is "yes," theirs is "no." I share Kelly's belief that the Amish example of slow adoption is instructive.

He offers four observations he has made about the Amish:

1. They are selective. They know how to say "no" and are not afraid to refuse new things. They ban more than they adopt.
2. They evaluate new things by experience instead of by theory. They let the early adopters get their jollies by pioneering new stuff under watchful eyes.
3. They have criteria that help them make their choices: technologies must enhance family and community and distance them from the outside world.
4. The choices are not individual, but communal. The community shapes and enforces technological direction.

These approaches work for the Amish, but can they work for the rest of us? Kelly isn't sure. As a culture, we are fast and furious adopters, lacking in the skill of abstaining.

"To fulfill the Amish model we'd have to get better at relinquishing as a group. Social relinquishing. Not merely a large number (as in a movement) but a giving up that relies on mutual support," writes Kelly.

Disconnecting Is Not Enough

Many people are picking up the "disconnectionist" banner. Even the likes of Arianna Huffington are urging us to unplug — for an hour, a day, a week. Unfortunately, it's often so we can resume our usual activities with even more vigor. After passing out in her office, Huffington started tackling the issue of overwork and constant stress (especially for women) on a part of her site called The Third Metric, which covers wellness topics, including how important a digital detox is for getting there.

"How can we support each other, as women, to actually create breathing spaces in our lives. To end this addiction. To take care of ourselves at the same time as we fulfill our obligations. That was what really led to The Third Metric," says Huffington. And I celebrate this. The problem is *Huffington Post* continues to contribute to the breakneck speed of publishing and consumption online. Sadly, there is something incongruent here.

Arianna is on to something. Perhaps we will see a new kind of *Huffington Post* yet.

Composing a Life

I am often asked, after my 31 days were up, how I returned to technology. In honesty, it was with fear and trepidation: I was nervous to reengage online. I didn't want to fall into old habits. Like reintroducing food after a water fast, I had to take it slow. First, I began by checking off my to-do list: unsubscribing from email clutter, implementing my 1x/day email rule. Ironically, at the suggestion of a writing teacher, I signed up for Twitter half a year *after* the experiment in an effort to promote ideas. The truth is, having three very small kids has helped keep me offline a lot: *I already need three extra arms.* Some days, usually when I am feeling a bit blue, extra rundown, I check Facebook 16 times. Like a painting, I'm a work-in-progress. I can choose to beat or pick myself up. I aim for the latter.

The great work for us all is composing a life. To do so, we need enough space to develop a clearer version of the life we want and enough imagination to envision an alternative to the way we are living.

There is an urgency to this. We are all willing adopters in this technological arms race. As more and more of us choose to pay more mind to screens than to the faces of other people, we have myriad problems on our doorstep: mistrust, addiction, unparalleled narcissism, decreased recall and critical thinking, diminishing intimacy and commitment to neighbors, an insatiable hunger for amusement (especially among the young).

As any parenting expert will tell you, it is within understood parameters that children thrive. We are no different. It is in setting limits and recognizing our own limitations, that we all — child and adult alike — experience a growing sense of freedom. Discipline leads to our joy.

4. Be Gentle with Yourself.

Perhaps the best advice comes to us from Ralph Waldo Emerson when he wrote: "Finish each day and be done with it. You have done what you could. Some blunders and absurdities no doubt crept in; forget them as soon as you can. Tomorrow is a new day. You shall begin it serenely and with too high a spirit to be encumbered with your old nonsense." Some blunders and absurdities have crept in, it is done, *let them go.* Now, may forgiveness and fluidity guide us as we chart a new course.

Follow the Way of Peace

I don't look lonely. My life is full. I have a gaggle of kids, a partner. I have seven siblings, and in-laws and parents and nephews and nieces. I know many neighbors on my block. Yet I have felt what Jean Vanier calls "a faint dis-ease, an inner dissatisfaction, a restlessness of the heart." (*Becoming Human,* "Loneliness" p 7) I have felt it deeply, and I have felt it grow online.

But abstaining from the Internet won't save us from our loneliness, from our restlessness and dissatisfaction. No, because, says Vanier, loneliness comes at any time. "It comes in times of sickness or when friends are absent; it comes during sleepless nights when the heart is heavy, during times of failure at work or in relationships; it comes when we lose trust in ourselves and in others."

We don't always feel our loneliness. "When people are physically well, performing creatively, successful in their lives, loneliness seems absent," Vanier continues. "But I believe that loneliness is something essential to human nature; it can only be covered over, it can never go away. Loneliness is part of being human, because there is nothing in existence that can completely fulfill the needs of the human heart."

Loneliness can be a source of creative energy, pushing us to create new things, walk new paths in order to seek truth and justice in the world. More often, though, loneliness shows itself in less positive ways. "It can push us to escapes and addictions in the need to forget our inner pain and emptiness." (*Becoming Human,* p 8)

Escapes and addictions may be where we find ourselves, but in composing a life, in letting our yeses be yeses and our noes be noes, a sense of clarity and lightness floods in. There can be self-forgetfulness because we are not worried about posting and sharing and tweeting everything into the ether but, instead, delighting our loved ones with our presence, our anecdotes, our ideas. There can be peace because we have made our choice.

I love what is happening on the Internet. I do. I love the ideas, the confessions, the livelihoods, but what I find there means nothing, *nothing,* if I don't take what's found there and give it legs. And to begin to give these things legs, I must stay offline *for a while* because my mind needs to dig into the thinking, needs to suss out the bits that don't quite make sense, and I need to make a plan — the list, the blueprint, the intention — and then I need to stick with it. And the Internet can help with that. It can help. It is a helper. It is not a be-er. It is not a do-er. No, those parts are up to me.

And, though I try to be honest, try to share things online that help, though I try not to draw too much attention to myself, what I put there is still not me. Not really. Because I've got skin on. And that skin, it matters. Because I was born with it. I didn't pick it; I've just got to grow into it. And, y'know, even though it's more spotty and less taut than it once was, I think it's getting more beautiful. And, you know what? I think yours is, too.

Conclusion

This Will Be Joy

I have written a cautionary book in hopeful expectation that time remains for corrective actions.

— Jane Jacobs

With our time and presence we give love. Simple.

— Kim John Payne

MY FRIEND CHRIS LIVES A BEAUTIFUL LIFE.

It's late winter when I spot him a block from Main and Hastings — Vancouver's seediest intersection. He's standing on the sidewalk, one hand balancing his bike, another reaching out to a fellow holding a handwritten sign, his not-yet-three-year-old wedding band glistening in the February sun.

When I mention it a few weeks later, he tells me he seeks out good conversation wherever he can find it, every day. "I am so boring," he says, "and other people are so interesting." Others might disagree. A regular guest host on CBC's Radio 3 for the last six years, Chris has interviewed some of the world's most interesting musical acts, including Berlin-based electro-pop diva Peaches and K-os. He is also a sought-after solo performer — Chris-A-Riffic — in his own right.

"He's a spectacle to see, with heartfelt gospel songs and mandatory crowd singing . . . he's at his excitable best when his face is red with spirit," reads one review.

> "I find that if you just sit in one spot, people want to unload what's on their minds."

Chris begins most days by cycling to a coffee shop, writing a little bit, and hoping he runs into someone to have a conversation with. The aptly-named Our Town Cafe is a regular haunt where he finds lots of opportunities for conversation. "I know a lot of the people who work there. There are music types and art types. I like to sit and make myself available to talk. It's not that hard for these moments to come up."

"I was sitting beside this guy at a coffee shop yesterday and he looked a bit rough. I think most people need someone to talk to. This is such a closed-off town and it seems like people are trying so hard to keep it together. I find that if you just sit in one spot, people want to unload what's on their minds. I have a circle of musical friends who I see at shows, but there are always these people who are not that adjusted socially. I love to talk with them the most, because they have a lot to say."

The truth is, people love Chris. He has the uncanny ability to set people at ease. The day we speak, he's begun the day the way he always does: spending a lovely hour with his drummer and costume-maker wife, Allison. Then for a morning spin on his bike, and onto the cafe for writing and talking. Later he'll teach piano for the lion's share of the day (his main occupation since the age of 17, and one he loves dearly) before performing that night at Music Waste, a local music festival.

Chris is in the midst of recording a four-song, seven-inch independently-released vinyl album at the small Presbyterian church where he's attended and played piano for most of his life.

"My dream job would be to do what I'm doing. I love to teach; I love to play music more than I ever have. My wife got me a piano for my birthday a couple of years ago and playing it is the best thing. I have never yet had to choose between work and my music."

Though he has tour plans in the works, Chris is not self-promoting. "I don't know where the line from confidence to arrogance is. It's a slippery slope from confidence to *I am wearing a golden crown . . . and purple robes . . .*

and you must address me as Emperor. . . . I am an unmover and unshaker. I am like concrete. I just write songs and like to play."

Meaning in Limitation

When Jean Vanier sat down to write his Massey Series lectures, he intended to title them "From Chaos to Life." In the process of writing them, however, he felt they should be called something different, settling on the theme of *Becoming Human.*

In the lecture he writes: "Is this not the life undertaking of us all, to become human? It can be a long and sometimes painful process. It involves a growth to freedom, an opening up of our hearts to others, no longer hiding behind masks or behind the walls of fear and prejudice. It means discovering humanity."

It's the truth of monogamy: we find meaning because there are limitations. As humans we find more meaning in limited connections, limited on- and off-line intake and output. To find this meaning, we must move from chaos to life, toward something simpler, together championing one another toward humanness.

> "Lasting change happens when people see for themselves that a different way of living is more fulfilling than their present one."
> — Eknath Easwaran

To Have and to Hold

So, lately you've found yourself pulling out old vinyl records, thumbing through family photo albums and dusting off old stationery to craft handwritten notes. You are not alone. In fact, you are in very good company. Seventy-nine percent of adults agree that oftentimes they miss having memories in physical form, such as photos, letters or books. In the ever-changing digital world that we live in, we find ourselves placing increasing value on the moments away from a screen. We are adapting. As our culture gets saturated by the Web, more and more of us are craving offline experiences, turning to the physical and tactile.

When I began writing this book, I felt the doom and gloom of the Internet, the unease about our modern world. I felt it to my core. But on the other side of the research, the conversations, reflection and writing, I see a different future on the horizon. Smartphones and social media are our modern-day spice and sugar. Imagine having your food, for the first time, enlivened by these flavors. Of course, we'd lather it on. But, eventually we'd settle in, aiming for more nuance and restraint. That's what I see beginning to happen in our use of the Web.

But restraint will not develop without open eyes and hard-won battles. The battle every day to put down the phone and look up. The battle every month to abstain from buying a new gadget when the old one works perfectly well. The battle to prioritize the people in front of us ahead of the phone in our hands. The battle to carve out time to spend with the lonely, sick and vulnerable in our lives.

This is the future I dream of: one where each of us thoughtfully engages online, putting people first and using the Internet for what it is — a tool.

The future is not an inevitability. It is up to us. In choosing to be alone less, and together more; in choosing simpler moments over complicated entanglements; in drawing close, instead of up and away; in doing these things, we are, ultimately, choosing vulnerability, and in it, together, we are becoming human.

Only love and what love forms, endures.

Jean Vanier concludes:

> "To be human means to remain connected to our humanness and to reality. It means to abandon the loneliness of being closed up in illusions, dreams, and ideologies, frightened of reality, and to choose to move forward in connectedness. To be human is to accept ourselves just as we are, with our own history, and to accept others as they are.
>
> To be human means to accept history as it is and to work, without fear, toward greater openness, greater understanding, and a greater love for others. To be human is to not be crushed by reality, or to be angry about it or to hammer it into what we think it is or should be, but to commit ourselves as individuals, and as a species, to an evolution that will be good for us all."

Our delights are signposts on the journey to life. The word delight is rooted in the word *delectare,* "to entice or allure away." Our true delights, the most momentous, life-giving experiences of our lives, entice us down the best path.

It is joy that leads us home.

The Ones Who Love Us Most

When I embarked on my offline experiment, I didn't want to go it alone. So I called one of my oldest friends, Marisa Ducklow, and asked if she was up for an adventure. With a teaching career and two little boys in tow, she somehow found the time to open mail, scan letters and post her thoughts on the Letters from a Luddite blog as envelopes trickled in in a curious order. One day letter number 14 arrived; the next day, number 3. And together, through the technology of the typewriter, the telephone, the scanner and the laptop, the journey brought two friends and an online audience together.

"I find myself sometimes staring at the screen wondering what exactly I am looking for here," wrote Marisa on day 17. "As if there is something, somewhere on this vast Internet, that will somehow fill the void or numb the feeling or enlighten my bored mind. Maybe if I Google the meaning of life or somehow stumble across the perfect, most inspiring blog, then suddenly everything will fall into place!

"But really what I need in those moments are all the friends who have moved far away, and who knew me all those years ago when I wasn't a mother or a teacher or a wife. Sometimes I just need to sit with them and remember when we would sit on the roof and drink tea and look at the stars. Or be reminded of when I woke up in the rain, yet in my bed, on one of my first nights as Christina's roomie. Or the time I came home and she was sitting on her bed sobbing amongst a huge pile of clothes on the floor. I was so astonished to see anything out of place that I didn't know if I should ask her what happened to her CLOTHES or what happened to her, and in the end I don't think I asked anything at all, just sat there and waited for the story to unfold.

"What I really need in these moments isn't an email or even a phone call, I need a jet plane to fly me off to Toronto or Montreal or Calgary or Nashville! But you know what seems to be the next best thing? A letter. A piece of frail paper fed into my Olympia deLuxe, typed out from the heart, and put in the mail to be received in 2 days or maybe 17 days, depending on the weather, the distance, and the mood of the postal service that week. And when that dear friend gets my letter in the mail, she might be changing diapers or drowning in laundry or sitting staring at her own screen wondering what she needs, and there will be a glimpse of the past, a note from a faraway friend who misses her dearly and still remembers the little things we did together."

In his book *The Courage to Create,* Dr. Rollo May, one of the most influential American psychologists of the 20th century, compiled a series of meditations, both wise and hopeful, on the future of mankind. It was 1975:

> "A common practice in our day is to avoid working up the courage required for authentic intimacy . . . Authentic social courage requires intimacy of the personality simultaneously. Only by doing this can one overcome personal alienation. No wonder the meeting of a new person brings a throb of anxiety as well as the joy of expectation; and as we go deeper into the relationship, new depth is marked by some new joy and new anxiety."

Life in all its fragility and beauty, in its imminent vulnerability, is what makes us human. And it is this that keeps us alive.

Recommended Reading

Alone Together: Why We Expect More from Technology and Less from Each Other, by Sherry Turkle (Basic Books, 2011)

Becoming Human, by Jean Vanier (Anansi, 1998)

Better Off: Flipping the Switch on Technology, by Eric Brende (Harper Perennial, 2005)

Daring Greatly, by Brené Brown (Gotham Books, 2012)

End of Absence, by Michael Harris (HarperCollins, 2014)

Hands Free Mama, by Rachel Macy Stafford (Zondervan, 2014)

Lightweb Darkweb: Three Reasons to Reform Social Media Be4 It Re-Forms Us, by Raffi Cavoukian (Homeland Press, 2013)

Quotidian Mysteries, by Kathleen Norris (Paulist Press, 1998)

Seven Vices of the Virtual Life, by Dr. Read Schuchardt (InterVarsity Press, forthcoming, 2015)

"Simple Thoughts for Engaging Technology," by Aiden Enns (*Geez Magazine,* Issue 20, Winter 2010)

Simpler Living, Compassionate Life, by Michael Schut, ed. (Living the Good News, 1999)

Slow is Beautiful, by Cecile Andrews (New Society Publishers, 2006)

Technology and the Character of Contemporary Life, by Albert Borgmann (University of Chicago Press, 1987)

The Shallows: What the Internet is Doing to Our Brains, by Nicholas Carr (W.W. Norton & Company, 2011)

What are People For? by Wendell Berry (North Point Press, 1990)

Bibliography

"Eye contact detection in humans from birth," www.pnas.org/content/99/14/9602.full Accessed August 1, 2014.

Blakemore, Sarah-Jayne, Joel Winston and Uta Frith, "Social Cognitive Neuroscience: Where Are We Heading?" www.icn.ucl.ac.uk/sblakemore/SJ_papers/BlaWinFri_Social_TICS04.pdf Accessed October 15, 2014.

Vanier, Jean, *Becoming Human,* House of Anansi Press, 1998.

Borgmann, Albert, *Technology and the Character of Contemporary Life: A Philosophical Inquiry,* University of Chicago Press, 1987.

Payne, Kim John and Lisa M. Ross, *Simplicity Parenting: Using the Extraordinary Power of Less to Raise Calmer, Happier, and More Secure Kids,* Ballentine Books, 2010.

"An interview/dialogue with Albert Borgmann and N. Katherine Hayles on humans and machines," press.uchicago.edu/Misc/Chicago/borghayl.html Accessed October 15, 2014.

"How Sitting All Day is Damaging Your Body and How You Can Counteract It," lifehacker.com/5879536/how-sitting-all-day-is-damaging-your-body-and-how-you-can-counteract-it Accessed October 15, 2014.

"Living in a World with No Off Switch," lecture by Dr. Read Schuchardt deliveredy at College Church, Wheaton, IL, 2008.

Thomas, Trudelle, *Spirituality in the Mother Zone: Staying Centered, Finding God,* Paulist Press, 2005.

Kelly, Kevin, "Amish Hackers," kk.org/thetechnium/2009/02/amish-hackers-a/ Accessed October 15, 2014.

Nerone, John and Kevin G. Barnhurst, "News Forms and the Media Environment," *Media, Culture & Society,* 2003.

Barnhurst, Kevin G. and John Nerone, *The Form of News: A History,* Guilford Press, 2002.

Bilton, Nick, "The 30-Year-Old Macintosh and a Lost Conversation With Steve Jobs," bits.blogs.nytimes.com/2014/01/24/the-30-year-old-macintosh-and-a-lost-conversation-with-steve-jobs/?ref=technology Accessed October 15, 2014.

Frances, MD, Allan, "DSM5 is a Guide, Not a Bible — Ignore its Ten Worst Changes," www.psychologytoday.com/blog/dsm5-in-distress/201212/dsm-5-is-guide-not-bible-ignore-its-ten-worst-changes Accessed October 15, 2014.

"Fifty Essential Mobile Marketing Facts," in forbes.com www.forbes.com/sites/cherylsnappconner/2013/11/12/fifty-essential-mobile-marketing-facts/ Accessed October 15, 2014.

Snow, Shane, "What Inspires Me: Flawed People and Underdogs," contently.net/2014/01/27/writers-desk/what-inspires-me-flawed-people-and-underdogs/ Accessed October 15, 2014.

Rainie, Lee and Barry Wellman, *Networked: The New Social Operating System,* The MIT Press, 2012.

Bruneau, Emile G. and Rebecca Saxe, "The power of being heard: The benefits of 'perspective-giving' in the context of intergroup conflict," *Journal of Experimental Social Psychology,* Vol. 48, Issue 4.

Waldrop, M. Mitchell, *The Dream Machine: J.C.R. Licklider and the Revolution That Made Computing Personal,* Viking Penguin, 2001.

"Places we don't want to go: Sherry Turkle at TED2012," blog.ted.com/2012/03/01/places-we-dont-want-to-go-sherry-turkle-at-ted2012/

Doyle Melton, Glennon, "Share This With All the Schools, Please," momastery.com/blog/2014/01/30/share-schools/ Accessed October 12, 2014.

"Albert Borgmann on Naming Technology: An Interview," by David Wood. www.religion-online.org/showarticle.asp?title=2901 Accessed October 14, 2014.

Vandehey, Jeremy, medium.com/p/15308056cfae

Carr, Nicholas, *The Shallows: What the Internet Is Doing to Our Brains,* W. W. Norton & Company, 2011.

"This is Your Brain on Jane Austen, and Stanford Researchers are Taking Notes," by Goldman, Corrie, 2012. news.stanford.edu/news/2012/september/austen-reading-fmri-090712.html Accessed October 15, 2014.

"Attention Spans Have Dropped from 12 Minutes to 5 Minutes — How Social Media is Ruining Our Minds," [Infographic] socialtimes.com/attention-spans-have-dropped-from-12-minutes-to-5-seconds-how-social-media-is-ruining-our-minds-infographic_b86479

Schut, Michael, *Simpler Living, Compassionate Life: A Christian Perspective,* Morehouse Publishing Company, 2009.

Krashinsky, Susan, "Marketing matters: The 'small' problem with mobile ads," in *The Globe and Mail,* June 29, 2013. m.theglobeandmail.com/report-on-business/industry-news/marketing/marketing-matters-the-small-problem-with-mobile-ads/article12894825/?service=mobile Accessed October 10, 2014.

Gallagher, Winifred, *New: Understanding Our Need for Novelty and Change,* Penguin Books, 2013.

"Ultra-wired South Korea battles smartphone addiction," in nydailynews.com www.nydailynews.com/life-style/health/south-korea-battles-smartphone-addiction-article-1.1387062#ixzz2vrmU9J1j Accessed October 1, 2014.

Brende, Eric, *Better Off: Flipping the Switch on Technology,* Harper Perennial, 2005.

Jaslow, Ryan, "Internet addiction changes brain similar to cocaine: Study," in cbsnews.com www.cbsnews.com/news/Internet-addiction-changes-brain-similar-to-cocaine-study/ Accessed October 10, 2014.

Stren, Olivia, "Rebel Rebel," in *Toronto Life* www.torontolife.com/informer/random-stuff-informer/2007/01/01/rebel-rebel-george-stroumboulopoulos Accessed October 1, 2014.

Thurston, Baratunde, "#UNPLUG: Baratunde Thurston left the internet for 25 days, and you should, too," in fastcompany.com, www.fastcompany.com/3012521/unplug/baratunde-thurston-leaves-the-internet Accessed October 15, 2014.

Berry, Wendell, *What are People For?* North Point Press, 1990.

"Japan to launch Internet "fasting" camps," in nydailynews.com www.nydailynews.com/life-style/japan-launch-internet-fasting-camps-article-1.1440483#ixzz2vrjCi6dv Accessed October 1, 2014.

"Hikikomori: Why are so many Japanese men refusing to leave their rooms?" in bbc.com www.bbc.com/news/magazine-23182523 Accessed October 1, 2014.

"Technology's Man Problem," in nytimes.com www.nytimes.com/2014/04/06/technology/technologys-man-problem.html Accessed October 1, 2014.

Jacobs, Alan, *The Narnian: The Life and Imagination of C.S. Lewis,* Harper Collins, 2005.

Brown, Brené, *Daring Greatly: How the Courage to be Vulnerable Transforms the Way We Live, Love, Parent and Lead,* Gotham, 2012.

"Phone Sex: Using our smartphones from the shower to the sack," in zdnet.com www.zdnet.com/phone-sex-using-our-smartphones-from-the-shower-to-the-sack-7000017935/ Accessed October 1, 2014.

Kathleen Norris, *The Quotidian Mysteries: Laundry, Liturgy and "Women's Work",* Paulist Press, 2004.

"Claiming an Education, speech by Adrienne Rich, delivered at Douglass College, 1977. isites.harvard.edu/fs/docs/icb.topic469725.files/Rich-Claiming%20an%20Education-1.pdf Accessed October 14, 2014.

"04-Lent: Reclaim the Sensory," in geezmagazine.org www.geezmagazine.org/magazine/article/04-lent-reclaim-the-sensory/ Accessed October 6, 2014.

Gallagher, Winifred, *New: Understanding Our Need for Novelty and Change,* Penguin Books, 2011.

"Scientists Fear Day Computers Become Smarter Than Humans," in foxnews.com www.foxnews.com/story/2007/09/12/scientists-fear-day-computers-become-smarter-than-humans/ Accessed October 6, 2014.

"Are the robots about to rise? Google's new director of engineering thinks so...", in theguardian.com www.theguardian.com/technology/2014/feb/22/robots-google-ray-kurzweil-terminator-singularity-artificial-intelligence Accessed October 6, 2014

"Ray Kurzweil's Plan: Never Die," in wired.com archive.wired.com/culture/lifestyle/news/2002/11/56448 Accessed October 15, 2014.

"The Inside Story of Oculus Rift and How Virtual Reality Became Reality," in wired.com www.wired.com/2014/05/oculus-rift-4/ Accessed October 14, 2014.

Cavoukian, Raffi, *Lightweb Darkweb: Three Reasons to Reform Social Media Be4 it Re-Forms Us,* Homeland Press, 2013.

"Sedentary Behaviours," in activehealthykids.ca www.activehealthykids.ca/ReportCard/SedentaryBehaviours.aspx Accessed October 14, 2014.

"Pediatricians call for limits on kids' screen time," in reuters.com www.reuters.com/article/2013/10/28/us-pediatricians-kids-screen-idUSBRE99R0MJ20131028 Accessed October 15, 2014.

Cain, Susan, *Quiet: The Power of Introverts in a World that Won't Stop Talking,* Crown Publishing Group, 2013.

"Is the Internet hurting children?" in cnn.com www.cnn.com/2012/05/21/opinion/clinton-steyer-internet-kids/ Accesed October 15, 2014.

Elkind, David, *The Hurried Child: Growing Up Too Fast Too Soon — 25th Anniversary Edition,* Da Capo Press, 2006.

"Smartphone Addiction," in psychologytoday.com www.psychologytoday.com/blog/reading-between-the-headlines/201307/smart-phone-addiction Accessed October 15, 2014.

"Cells That Read Minds," in nytimes.com www.nytimes.com/2006/01/10/science/10mirr.html?pagewanted=all Accessed October 15, 2014.

"This is Your Brain on Mobile: A critique of destructive smartphone habits by someone who makes a living off of them," in medium.com medium.com/@jgvandehey/this-is-your-brain-on-mobile-15308056cfae Accessed October 15, 2014.

"Five Myths About Young People and Social Media," in psychologytoday.com www.psychologytoday.com/blog/freedom-learn/201402/five-myths-about-young-people-and-social-media Accessed October 15, 2014.

"Family and Technology: The #1 Rule For All Parents," in huffingtonpost.ca www.huffingtonpost.com/dr-karyn-gordon/family-and-technology-1-rule_b_3069523.html Accessed October 15, 2014.

"Our economic future depends on early investment in kids" in sfgate.com www.sfgate.com/opinion/article/Our-economic-future-depends-on-early-investments-5194170.php Accessed October 14, 2014.

"The Flight from Conversation," in nytimes.com www.nytimes.com/2012/04/22/opinion/sunday/the-flight-from-conversation.html?pagewanted=1&_r=3 Accessed October 14, 2014.

"The End of Childhood," in adbusters.org www.adbusters.org/magazine/78/end_of_childhood.html Accessed October 14, 2014.

Stafford, Rachel Mary, *Hands Free Mama: A Guide to Putting Down the Phone, Burning the To-Do List, and Letting Go of Perfection to Grasp What Really Matters!*, Zondervan Trade Books, 2013.

"Gamers succeed where scientists fail: Molecular structure of retrovirus enzyme solved, doors open to new AIDS drug design," in sciencedaily.com www.sciencedaily.com/releases/2011/09/110918144955.htm Accessed October 14, 2014.

"Why We All Need to Belong to Someone," in psychologytoday.com www.psychologytoday.com/blog/in-the-name-love/201403/why-we-all-need-belong-someone Accessed October 15, 2014.

"12 Months of Wellness," in alive.com interactive.alive.com/september-2013/12-months-of-wellness-2/# Accessed October 15, 2014.

About the Author

CHRISTINA CROOK is a writer whose features on art, culture and technology have appeared in UPPERCASE, CBC.ca, *Vancouver Magazine, Today's Parent, Geez, Faith Today* and the *Literary Review of Canada.*

In 2012 she disabled the data on her smartphone, turned off her email and said goodbye to the Internet for 31 days. She chronicled the journey with daily letters which were mailed, then posted to a blog by her friend Marisa, creating a conversation between friends and open to the world at large. This experience, chronicled as the project, *Letters from a Luddite,* garnered international media attention and fueled Christina's passion for exploring the intersection of technology, relationships and joy.

CREDIT BETTINA BOGAR

Crook has worked as a journalist and communications professional for some of Canada's most recognized organizations, including the Canadian Broadcasting Corporation and Rogers Digital Media. She is a graduate of the Simon Fraser University School of Communication and her TEDx talk, "Letting Go of Technology: Pursuing a People-focused Future," was presented as part of the 2013 Global TEDWomen conference.

Crook is a proud sister, wife and a mother of three. She and her young family recently traded the seaside views of Bowen Island, British Columbia for the banks of Toronto's Humber River where they attend Grace Toronto Church.

New Society Publishers

ENVIRONMENTAL BENEFITS STATEMENT

New Society Publishers has chosen to produce this book on recycled paper made with **100% post consumer waste,** processed chlorine free, and old growth free.

For every 5,000 books printed, New Society saves the following resources:[1]

23	Trees
2,121	Pounds of Solid Waste
2,333	Gallons of Water
3.044	Kilowatt Hours of Electricity
3,855	Pounds of Greenhouse Gases
17	Pounds of HAPs, VOCs, and AOX Combined
6	Cubic Yards of Landfill Space

[1]Environmental benefits are calculated based on research done by the Environmental Defense Fund and other members of the Paper Task Force who study the environmental impacts of the paper industry.